TELEPHONY:
Today and Tomorrow

PRENTICE-HALL SERIES IN
DATA PROCESSING MANAGEMENT

Leonard Krauss, Editor

TELEPHONY:
Today and Tomorrow

Dimitris N. Chorafas

PRENTICE-HALL, INC., *Englewood Cliffs, N.J.* 07632

4/16/85

Library of Congress Cataloging in Publication Data

Chorafas, Dimitris N. (date)
   Telephony: today and tomorrow.

   Includes index.
   1. Telecommunication—Technological
innovations.   2. Telephone—Technological innovations.
I. Title
TK5101.C496   1984   621.385   83-3271
ISBN 0-13-902700-9

Editorial/production supervision: Lori Opre
Cover design: Ben Santora
Manufacturing Buyer: Gordon Osbourne

Printed in the United States of America

10 9 8 7 6 5 4 3

ISBN 0-13-902700-9

Prentice-Hall International, Inc., *London*
Prentice-Hall of Australia Pty. Limited, *Sydney*
Editora Prentice-Hall do Brasil, Ltda., *Rio de Janeiro*
Prentice-Hall Canada Inc., *Toronto*
Prentice-Hall of India Private Limited, *New Delhi*
Prentice-Hall of Japan, Inc., *Tokyo*
Prentice-Hall of Southeast Asia Pte. Ltd., *Singapore*
Whitehall Books Limited, *Wellington, New Zealand*

CONTENTS

PREFACE

We are beginning to think of telecommunications as a tool of business rather than as an ancillary service. Telephony is no longer subordinate to a paper-based office system but is rather the pivot point around which other systems start to develop.

In the business of telephony, the players are changing and will continue to change, since the stakes are high. The new product line of the telephone companies (telcos) will no longer be limited to voice systems, with a private branch exchange (PBX) as the exception. It will: include computer-based telephone systems; offer voice mail; support text and data entry; incorporate intelligent terminals, copiers, and word processors; and use the latest technologies—optical fibers and satellites.

Local and long-haul networks for integrating voice, text, data, and image will take a *systems approach*. The marketplace will see major companies getting into the game—through separate subsidiaries or joint ventures, as in the case of AT&T, Xerox, and IBM; and through acquisition, as in the case of GTE and Telenet.

With computers and communications merging into a new and broad discipline, competition will center on services that promise a total *compunications* (*compu*ters and commu*nications*) solution to the end user. The requirements of the marketplace, advances in technology, and a rapidly changing regulatory picture are all converging to bring about this change—and with a great impact on society.

The emerging specialized value-added networks (VAN) will be employing substantially more intelligence to provide an increasing number of supports to users:

- Message store and forward
- Analog/digital conversion
- Compression capabilities for data, text, voice, and image
- Improved error detection and correction
- Utility-type mainframe support
- User programmability and off-the-shelf packages
- Statistical collection and reporting
- Remote diagnostics
- Network maintenance and management

How can we foresee the highlights of this development and plan for the coming evolution?

This is the subject of the book the reader has in hand. Part One is an introduction to the voice network. What do we mean when we talk of telecommunications? Are we still bound to concepts developed a century ago? Are there coming developments that can bring POTS (the plain old telephone service) into a new era? The answer is *yes*; and it is documented.

The telephone was invented by Alexander Graham Bell in 1876 and underwent little change until nearly a quarter-century later, when the Strowger exchange was developed. Further improvements have been comparatively slow—until recently. During the last two decades, telephone evolution has been accelerating. The telephone is an interactive message service, and we shall be looking at it in this sense.

End-to-end communications have come into clearer perspective. Store and forward abilities eliminate the necessity of interconnecting the sender to the receiver (thus permitting voice mail). Digital technology has invaded the analog bastion of human voice, and bit error rates have dropped dramatically; this has opened up the lines to new possibilities of usage. A direct-dialing technology has been developed and the numbering system extended internationally, so that we can now call from New York to Sydney, Haifa, and Rome without human interface.

Tariffs do play a leading role in shaping these features, and the more we use telephone services the more we find the costs associated with them to be staggering. Technology opens up opportunities, but someone must foot the bill. As tariffs change—becoming *volume bound* rather than *range determined*—and as more devices are hooked up to the line, we call on new equipment *to expand the capacity*. This establishes a need for the PBX. Computer-based and programmable, the modern PBX is no kin to the switchboards of yesterday.

Part Two focuses on some of the new technologies. The chapter on microwave links explains the theory in a nutshell and is used as an aid to presenting fiber optics and satellites. Two chapters are devoted to optical fiber communications and some of their possible implementations, including components, interfacing, and systems engineering. Another chapter is dedicated to multitap coaxial cable and the types of networks it can support.

Three chapters treat the satellite systems: their all-digital, multimegabit, rooftop-antenna technologies; the information transfer services they can provide; the technological evolution toward large satellites prompted by the Space Shuttle; issues relating to antennas; and the technical and political problems inherent in these developments. Earth stations and system responsibilities for managing satellite systems are also emphasized.

Because experience leads to maturity, Part Three is dedicated to modern communications networks. First the role of communications in our society is critically examined: we use lines to transmit voice and data, and the radio broadcast for voice and images; we feel that office efficiency is suffocating under mountains of paper and irretrievable text references; and we look around for solutions. Somebody was bound to come up with the idea of integrating the various services.

We have finally realized that information services distinct for decades in fact complement one another and can be treated nicely on the same network if it has an appropriate architecture. But which goals should it serve? In which way will the messages be handled? What is meant by *session control*? Are there any standards? What do they imply?

The middle to late 1980s will present enormous growth opportunities to the common-carrier industry and to manufacturers of communications equipment for both local and long-haul networks. What functions will a local network be expected to perform? How will it be structured? Is it possible to have design alternatives?

The market for delivery of electronic information for both local area and long-haul networks will grow rapidly as soon as the end user can justify the cost of implementing new applications. Bell's Net 1 is used as an example of long-haul services because it is the most likely to set new standards.

The growth of the local area network (LAN) market is fueled by aggressive semiconductor manufacturers who are mass-producing low-cost communications components. When it becomes cost-effective to digitize in the telephone, an exploding market for high-capacity, digital, local and long-haul networks will result, providing the backbone for a new communications capability within the office and the home. Quite reasonably, these developments will have an impact on privacy. To protect our freedoms while benefiting from the new technologies, we need farsighted legislation. A social conscience is beginning to develop regarding the form that may be taken by the transborder data flow; we must be careful to avoid alienating the public. Accordingly, farsighted legislators should try to inform, convince, and regulate.

We should invest in our human capital through the right training; automating a clerical function does not necessarily improve productivity, service, or job interest. Success will ultimately depend on the human beings who use the system, and the better trained they are, the greater the results that we can expect from our investment.

It is not enough to recognize that information is a marketable product. We must also be in the forefront in heeding some of the educational, social, political, and economic implications of the technological revolution. Lessons can be learned from our master strokes, and even better lessons can be learned from our mistakes. Nothing great will ever be achieved without people—and people can achieve greatness only if they are determined.

Let me close by expressing my thanks to all who contributed to making this book successful: my colleagues, for their advice; the organizations I visited in my research, for their insight; and Eva-Maria Binder, for her drawings and typing.

DIMITRIS N. CHORAFAS

Valmer and *Vitznau*

Chapter 1

INTRODUCTION TO TELECOMMUNICATIONS

INTRODUCTION

There are currently three main classes of communication systems used in business: the telephone—by far the most frequently used communication method; the postal service—government, private couriers, and intercompany mail delivery networks which physically carry messages; electrical message systems—principally telegrams, telex, and datex, but also more recent electronic solutions such as teletex, interactive videotex, and the new types of facsimile.

The telegraph, dating back 150 years to the development of Morse code, is the earliest electrical transmission medium. Telex in Europe and TWX in the U.S. are post–World War II dial-up message switching networks originally designed to transmit binary signals (pulses) at the 50 to 70 baud level (cycles per second). Datex is a European dial-up network for transmitting binary signals first at 200, then at 1200 baud and above. All three principal systems are employed primarily for low-speed transmission of binary signals. They employ direct-current (dc) networks with voltages between 40 and 120 volts.

The telephone network, based on Bell's discovery, works on line-switching principles and uses lines at 1200 baud (3000 to 3400 Hz). These are generally known as "voice grade" lines. Leased circuits (dedi-

1

cated lines) allow quality specifications (conditioning) and offer higher speeds at a given transmission error rate than dial-up connections can afford. Presently, however, leased lines are costly.

After decades of practically unaltered supported services, telephony has begun taking on new functions. Publicly offered data communications networks assure high-quality service; value-added networks support a variety of useful functions; and even the classical voice circuits offer such novelties as the "national phone number." With this feature, persons dialing that number will be able to trace us wherever we are—unless we use the rejection option that won't acknowledge designated calling numbers. "Call screening" is very handy for fending off persistent callers. It will also be much easier to obtain mobile phones for "follow me"-type communications, although increased voice, text, data, and image traffic will risk overloading the frequency bands.

A possible carrier saturation may be avoided in various ways, such as the capacity transfer capabilities projected when the new generation of satellites is in service. With this, users can transfer their rented bandwidth from coast to coast as peaks in traffic follow the sun zones. Because satellites will support an impressive bandwidth, video conferencing will be not only technically but also economically feasible.

The answer to the question: "How *broad* is broadband?" can vary as a function of time, technology, and of our own requirements. An easy, though not technically precise way to remember is that broadband means a greater frequency range than our *current* requirements call for. In technical terms, a different answer should, however, be given.

Let's take as an example coaxial cable transmission within the perspective of a local area network (LAN) implementation. We speak of *baseband* solutions when the channel capacity is in the range between zero and 10 megabits per second (MBPS); of *broadband* when the usable range varies from, say, 5 MBPS to 400 MBPS. Satellite transponders work broadband.

The communications industry has participated in and benefited from technological progress. One example is the relative cost per circuit-mile for new terrestrial transmission systems as a function of the number of circuits carried. For paired cables with 10 circuits, the relative cost is over $200 per circuit-mile, but the investment cost for a waveguide transmission circuit can be well under $1 per mile when 100,000 or more circuits are installed.

The new facilities offered by the communications industry are sure to have an impact on office work. Modern office communications systems are increasingly computer-based and able to provide integrated facilities to process messages with little or no use of paper records. Typically, such messages are produced, transmitted, received, stored, and retrieved by workstations which are more productive than manual solutions and improve work quality.

Telecommunications is a major factor in this new environment. Typing, assembling text and data, copying, calculating, filing, and retrieving must share the clerk's time with covering telephone calls, transmitting, and expediting. But a significant part of upper-level management time is concentrated on communications-oriented tasks. Is there any way to improve them?

RETHINKING THE MESSAGE SERVICE

In the United States about 90 billion pieces of mail move annually. Some 10% is writer-to-reader correspondence, including 4 billion business letters. Interoffice correspondence accounts for between 12 and 15 billion letters, memos, and so on. In addition, some 220 to 250 billion telephone calls and 50 million telex transmissions are made annually.

Figures on the length of messages provide interesting statistics. An estimated

- 58% of management messages are of 1 page or less.
- 28% are of 2 or 4 pages.
- 14% are of 5 pages or more.

But electronic message systems carry much shorter texts. Of all messages transmitted, 73% contain fewer than 1000 characters and

- 45% go to an individual
- 35% go to more than one individual
- 18% go to groups
- 2% go to a combination of groups and individuals

Furthermore, statistics on office work indicate that

- 40% of the manager's time is spent on mail, telephone, and business travel, with
- 12% to 35% spent on writing and reading.

Still another study identified the percentage of working time some professions spend to "push paper" (Figure 1-1). Statistics show that 550 kilograms of paper is printed each year for every person in America (a good share of this is computer output). One story claims that published technical articles are read, on the average, by six readers.

The computer, initially viewed as a means of thinning out the paper

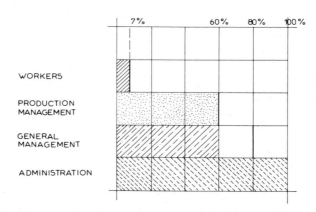

FIGURE 1-1   Percent of personnel "pushing paper" (in man-hours).

jungle, actually increased its density. It is now hoped that data communications, added to the range of electronic message systems in about 1960, will perform this reduction.

In general, electronic message systems have two primary types of message communication modes: hardcopy and softcopy. However, the telephone and postal services dwarf other message systems. If the telephone communication is excluded and only hardcopy intercompany and intracompany messages are considered, the following picture emerges.

The U.S. Postal Service (USPS) revenues break down to $5.5 billion from first-class mail and $5 billion from all other classes, including magazines, advertising, and packages. But this material must be delivered by hand, and communications revenues have already surpassed USPS's as a business-to-business message carrier of first-class mail (44% of $5.5 billion).

Looking at the electronic message system, we observe that its messages are normally of higher priority than those in other areas and that communications can be carried out by many different methods. One of the early ventures in electronic mail is the "Mailgram," a message-switching service jointly operated by the USPS and Western Union, but it still has a long way to go before it can compete with regular USPS rates or the new electronic mail products being developed.

One of the most significant new developments in interactive message systems was the implementation of "Interactive Videotex" (viewdata) in Great Britain in March 1979. A teletex (Electronic Mail) service is scheduled to operate when the telephone network is not used, from late at night to early in the morning.

If first-class mail can be delivered, so can third-class mail such as advertising. The reader not interested in that matter can easily identify and destroy it while reviewing the messages in his or her in-box. Otherwise, it can be displayed in color on a TV set.

Most important, voice storage and forward capability can be developed for use on the telephone and perhaps the TV network. The sender could direct a verbal message to the recipient's in-box. The latter could call his or her message storage from any telephone and get the message. Hotels might provide a means of coupling the telephone and the TV set to offer guests terminal capabilities for receiving and sending mail.

Merging message and voice facilities along the principles of packet switching will decouple sender and receiver, eliminating the need for both to be present. It will offer electronic speeds and geographic independence, and promote the use of computer technology to aid in composition, storage, retrieval, and reading.

Such solutions can operate within different environments:

- Exclusive station-to-station call
- Primary addressing, with copies to other stations
- Broadcasts, with messages sent from one station to a distribution list
- Teleconferencing (group participation)

But early experiences pinpoint several problems: the users have not always known what to do in time-out; we have not yet the expertise to help a user in distress; we have not learned how to saturate the topology without increasing costs; and as with all phases of office automation, we must immediately train the end user in all modifications.

Furthermore, while we are increasingly orienting our interest toward the wide range of text and data communications technology now makes feasible, we should not forget the vital role played by the voice services as we have known them during the last 100 years. Several issues are leaving their imprint in this sector, one of them being the landmark out-of-court settlement through which AT&T has agreed to give up the 22 Bell System operating companies that provide roughly 80% of the country's telephone service.

This restructuring will dramatically change AT&T as a company—it will also affect the average business user. Prior to this spin-off of a property valued at $80 billion, according to a company spokesman, the AT&T Long Lines' revenues subsidized the costs of local calling to the tune of 40 cents on the dollar. Without such subsidy, local rates can be expected to go up as the operating companies will be free to establish their charges (albeit within tariff constraints).

Two issues come immediately into perspective: First, the need for the telephone users (particularly the large ones) to organize so as to improve the cost/effectiveness of communications at large (voice communications in particular). Second, the requirement to redefine telephone usage by employees, tracing a thin red line between what is real business use and what is personal usage of the company's voice communications facilities.

There is indeed a certain confusion in business about telephone use as a fringe benefit. Many employees regard as such the use of a WATS. Similarly, they look at unlimited local calls as their prerogative. Management disagrees: WATS lines and intracity local calls are no longer billed at a flat rate. They cost an increasing amount of money.

There are some estimates that employee telephone abuse is costing American business $4 billion a year in work time and telephone charges. Better controls will help to bring this situation under control, but the real answer lies in

- employee education and
- capital investment in telephony which help bend the curve of steadily increasing service charges

In regard to employee education, we have to develop an awareness that the employee's best interest lies in the well-being and profitability of his or her firm. Regarding the second issue, vendors have responded with a range of hardware and software offerings in various price categories. We will review them most carefully when we talk of the Private Branch Exchanges.

COMPUTERS AND COMMUNICATIONS

During the last 10 years the evolution of equipment has been directed at meeting the developers' needs. What has been missing are procedural and end-user studies to help in projecting, evaluating, implementing, and maintaining future systems. For instance, wideband communications would allow a copier to produce a remote image at the same speed as a local copy. Wideband communications would be reserved for local copying during the day, multiplexed with receiving capabilities. At night, a single copier could receive a few thousand pages—data compression units can receive about 800 pages in 8 hours—and a facsimile transmission that now takes 6 minutes can be reduced to about 6 to 8 seconds using wideband.

The speed at which facsimile images can be carried should prompt electronic mail networks in the top 200 companies in the United States. By connecting major operating centers, these companies will be able to operate in a communications environment of unprecedented magnitude. Let's underline in this connection that facsimile is only one of the technical solutions for message exchange. Much electronic mail is not facsimile. Another broad area of message exchange is covered through the character-by-character asynchronous type of data communication characterizing the teletypewriters (TTY); still another through synchronous protocols, from BSC to packet switching.*

Thus, all told, the late 1970s have witnessed, and the early 1980s promise to extend, a rapid transition of computer-supported communications networks from experimental projects to full commercial development. Problems remain, however, particularly in flow control, congestion control, user access, and security.

Communications media have been designed primarily for telephone traffic, whether analog or digital. Since the late 1960s, a new technology, "packet switching," associates conventional communications techniques and computers to provide cheaper, more reliable service and to handle pieces of data directly. A significant contribution of packet switching is the radical change in network architecture and protocol.

Earlier network architectures were mainly of the inflexible star type, with terminals linked to the one central computer. They are bound to obsolescence, since such basic functions as transport facilities, terminal handling, and access methods are more or less intermingled. With star networks the protocols are asymmetrical, based on telephone line properties, and lack clear structure, since they mix terminal handling, application management, and computer constraints in ad hoc implementation.

New architectures provide a careful, layered approach by separating functions (protocols) into consistent, independent levels (Figure 1-2).

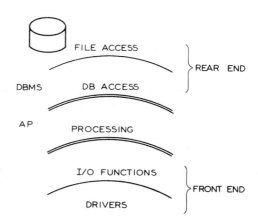

FIGURE 1-2 New architectures provide a layered approach, separating protocols into consistent levels independent of one another.

FILE ACCESS

DBMS   DB ACCESS

REAR END

AP   PROCESSING

I/O FUNCTIONS

DRIVERS

FRONT END

* See also D. N. Chorafas, *Data Communications for Distributed Information Systems*, Petrocelli Books, Princeton, NJ, 1980.

Thus various computers and terminals may be connected and share a common communications network which acts as a transparent data freeway. Terminals are connected to such a network through minicomputers, or they may interface directly if they can handle the protocol, typically the case with intelligent terminals.

The layered structure helps to make communications protocols independent of accidental properties encountered in text or data terminating equipment, transport facilities, and computer software. Assumptions about each layer properties are general enough to apply to existing and foreseeable networks.

In essence, a communications mechanism boils down to a limited number of very simple elements, but standards are very important. Standards in computer networks favor user interests. At higher levels in the network architecture, standards would be desirable for end-to-end protocols, virtual terminals, and command languages.

END-TO-END PROTOCOLS

End-to-end solutions necessarily involve flow control. The objective of a flow control scheme is to prevent overflow. Efficiency is also given more attention. In a layered architecture, flow control breaks down into two classes: end to end and congestion.

Simple mechanisms are sufficient to keep data flows under control (loss, duplication, out of sequence). Useful throughput and transit delay measurements can be conducted taking into account a number of variables: the packet length, the route followed inside the network, and the link control used on the various transmission lines.

The best choice depends on the requirements derived from the applications perspectives:

- What do we wish to do?
- What is the user's environment?

The first application of computer networks is to give terminals connected to them access to the services offered by the various computers and databases in the network. As it is not practical to incorporate procedures and terminal drivers suitable for all terminal models in each computer, one approach is to define a software interface corresponding to an imaginary terminal, known as the "virtual terminal."

Exchanges between computers and the network are then tailored to this standard interface. A real terminal is connected to the network through software which simulates a virtual terminal and is, of course, designed for the type of terminal concerned. In this way, compatibility can be obtained between nearly any computer and any terminal.

The connection of hosts is another issue of great importance. There are a variety of ways to interconnect host computers to a network: one is to implement all network protocols in the host system; another is to emulate in a special interface processor some terminals that the host system already supports, and then implement the network protocols in this interface.

Coordinating processes running on different hosts poses another major problem. Processes belonging to the same application share virtual common memory access areas. They are physically duplicated on the different hosts. The processes must be synchronized by means of primitives, and their performance must be measured.

Interfacing computer networks can also be a problem. Interfacing requires a well-defined and widely accepted host-to-host protocol. Transport stations located in each host must provide their users—programs, jobs, processes, and so on—with a transport function built on top of the communications functions.

Users (subscribers) exchange messages (memos, documents, reports, letters, or events) requiring services such as error control and flow control, and the problems we have outlined must be solved in order to satisfy user objectives. This was also the goal of the early star-type networks, but the increase in terminals and traffic volume, and the sensitivity to data errors and system failures, brought about a need for new network architectures.

STORE AND FORWARD MESSAGES

Store and forward messages are typically terminal-to-terminal or unsolicited host output text and data transmissions, stored on disk and routed to the destination. They involve a wider variety of destinations than does central memory-switched traffic, and require more extensive routing capabilities.

For central memory-switched traffic, routing functions are necessary to extract the destination logical identifier, determine the destination line and station tables, complete the control block for output with the appropriate routing data, and queue the control block for output.

Similar functions are necessary for disk-switched traffic, the principal ones being:

- *Edit:* extracting the routing data from the message content
- *Route:* using the routing data to determine the logical identifier and then the line and station tables
- *Tag build:* completing the control block and building additional control blocks as required for multiple deliveries

- *Queue:* queuing the control blocks, sending message acknowledgments, and writing logs as necessary

The queue function warrants additional comments. A typical queuing program writes the control block(s) to the appropriate output queue(s) on disk. It also updates the queue pointers retained in core, sends acknowledgments when required, sets the flag for output service action, and may perform logging. An output queue service program reads the control block entries from disk queues, reads the message segments from disk, and interfaces with the appropriate input–output routine to initiate output.

The store and forward process, developed after World War II for the message-switching telex network, together with packet switching, voice store, and forward capabilities, is sure to leave a deep imprint on the office environment.

VOICE MAIL

Designing for internal efficiency within the modern bank environment, we should refer to voice mail as implemented by a specially designed, computer-based private branch exchange (PBX). A PBX is a micro- or minicomputer-supported switch which can serve:

1. Line control functions
2. Voice/message input (store and forward)
3. Voice/message delivery
4. Dialogue communication
   a. Human–human
   b. Human–machine
   c. Machine–machine
5. Telex and teletex implementation
6. Pure data handling with X.25 interfaces
7. Facsimile exchange
8. Business for private viewdata
9. Multifunction workstations

The technology is available to assist an office worker with telephone, interphone, facsimile, teleprinter/telex, and data terminals. With the right equipment, people can do 50% of their office work at home, like Swiss watch artisans, and we can improve their performance at the workplace.

During the last three years, PBXs have taken on management and control features, a significant development from the past, when PBXs concentrated on switching inbound and outbound calls at a central site. Automatic route selection, call-detail recording, queuing, speedcalling, and integrated voice and data switching to reduce the need for wiring and modems are examples of the added features.

Still, as computer support becomes vital in PBX design, more sophisticated approaches can be taken: for instance, managing people's time by evaluating PBX data. The sales manager can take advantage of the record to manage the sales force. A salesperson's revenues, for instance, may be related to phone usage. Perhaps a salesperson spends 10 minutes on each call, whereas a successful seller averages only 4 minutes.

Recording employees' use of time on the job and accurately charging departments is a good method of pinning down costs and using people more productively. Our knowledge is meager and unsatisfactory unless we can measure it and express it in numbers. Any measurable factor in office operation can be automated to take advantage of the computer's ability to store and manipulate text and data. Properly directed, any job stream can be formatted and presented to the end user as an integral part of a management information system. In terms of information systems implementation, organizations expect their computer specialists to take the initiative in proposing new recording and control capabilities; one imaginative approach is the integration of voice, text, and data.

Communications experts pay a great deal of attention to the "voice mail systems" (VMS) actively under development at 3M/Electronic Communications Systems (ECS), AT&T, IBM, Northern Telecom, Rolm, Datapoint, and Wang. (Wang's version, the DVX or Digital Voice Exchange, was introduced to the market in late 1981.)

ECS's voice message system was the first to be announced (June 1980). Its function is that of an information processing system with electronic mail capabilities. As such, it is the missing link to the office of the future. With VMS, calls can originate from any tone-generating phone or from dial phones with tone-generating attachments.

Voice mail is a totally different approach from word processing (WP) and other automated office developments aimed at improving the productivity of secretaries and typists. Voice mail is aimed at the executive, and its goal is to do away with "telephone tag," one of the most frustrating office games: Mr. A calls Mr. B, who is away, so someone takes a callback message. When Mr. B returns the call, A is on another line, so B leaves a message. Mr. A calls again, but now B is in a meeting. And on and on.

Electronic mail does away with all this by turning spoken words into "digital mail." As any message switch, it divides into three parts:

1. The header (digital address)
2. The voice message
3. The trailer (special instructions: deliver tomorrow, redial in 1 hour, and so on)

Memos, sales orders, confirmations, reminders, and personal messages are examples of applications that are particularly appealing to managers reluctant to learn to use WP machines. Many managers spend three-fourths of their time talking to people, and voice mail can provide them with significant productivity enhancements.

The original voice response systems are a forerunner of this development. They were shown in trade shows since the mid 1960s, but took some time to get user acceptance. In 1974 the Philadelphia National Bank installed "Periphonics" touch-tone VRSs at the tellers' workstations, and BRED used a similar system in Paris. The same year, Rohr Industries successfully used Wavetek equipment for voice answerback. But such solutions remained isolated because most banks and industries used hardcopy or video units for their responses.

For whoever had the foresight to move away from the written word, visualization has largely carried the day over voice. Yet voice has its advantages because it exploits a different sense than that of the "crowded" eye. With this in mind, engineering is steadily searching for other channels. Tactual and pain stimuli have also been considered as alternatives.

At the current state of the art, voice input involves:

1. PAM-PCM conversion to make feasible transmission storage and playback
2. Simple transport with store and forward (short message, memo solution)
3. Text and data input (high technical requirements)
4. Evolution toward longer messages and voice signature (autopay)

Just the same, voice output can have a great impact on text and data processing at both the storage and reporting ends. It is no accident that a number of manufacturers (Texas Instruments, for example) have already produced equipment able to handle simple phrases. More complex units in development do not consider a fixed length. (Periphonics: about 70 words spoken with DEC 11/34 support.)

On the other hand, the principle of voice-to-voice communication stayed practically stagnant for over a century. Yet the need is pressing. A recent study found that roughly

- 75% of telephone calls do not reach the right person.

- 50% of the calls that are completed are one-way voice transmissions: requests, orders, notifications, replies to a request.

VMS will probably not replace two-way voice communications, but it will still save a manager's time. Once the caller's voice has been converted into digital form, the message can be

- Handled as if the information had been generated on WP equipment
- Sent to groups of people (distribution lists)
- Transmitted at night to a distant office with another voice-mail system for next-day delivery

Both private and public VMS service would appeal to users.

The 3M/ECS system now offered costs in the range $300,000 to $500,000, but an AT&T computer-based equipment test provides simple voice-mail storage and delivery services for as low as $15 monthly. However, 3M estimates that the "per voice message" cost of its equipment can be quite reasonable, taking its own headquarters as an example: on the basis of 15 messages daily at a $10 monthly charge to each user (for 1000 users) of VMS services, the cost is a mere 3 cents per message.

More significant than low cost, the super-PBX of this decade will be designed to serve as the nerve center of the automated office and handle all forms of digital communications: voice, data, text, and image. Software developments will make it feasible to pack more data into computer memory (to counterbalance the current memory requests of up to 300 times more storage capacity than WP input), and eventually voice delivery will assure the most natural form of communications—human speech—for text and data input.

By combining vision and sound technology, voice-mail revolutionizes communications perspectives while applying known tools—TV, PBX, telephony—in new systems.

Chapter 2

TELECOMMUNICATIONS SERVICES

INTRODUCTION

As used in Chapter 1, the word "network" was broadly defined to encompass not only voice and data transmission circuits (and lines) but also terminal equipment such as telephones, teletypewriters, and switching equipment (e.g., private branch exchanges and computers). New uses, such as data and facsimile, have only recently emerged in a service that has otherwise remained basically unchanged for 30 or 40 years, and as telecommunications ascends the corporate ladder, management will come to demand increasing professionalism of managers.

Several factors are causing the upsurge in telecommunications as a key management discipline: the advent of alternative suppliers of equipment and services; the increasing sophistication required for optimal choice; and the growing demonstration of benefits to management functions, such as marketing, production, and finance.

There is an increasing awareness of the role of telecommunications for corporate travel: moving information electronically without moving people and paper. As a result, corporate management is beginning to regard telecommunications less as a necessary evil and more as a vital management tool.

Telecommunications technology has been enriched by inventions in semiconductors, computers, and production engineering. The combination of computers and telecommunications technology has already borne fruit. Banks are an obvious example: no major financial institution today could contemplate running its business without online systems that not only serve clientele more efficiently, but also provide an invaluable marketing tool. Banks are consolidating credit and credit card information, and offering services to a far wider range of customers than the limited branch banking setup allowed previously. Airlines have long computerized their reservation systems as well as their arrival and departure information. Through point-of-sale terminals, the retail trade has moved toward online credit authorization and almost instant information on the sales of particular products, using that information to reorder and avoid both heavy inventory and out-of-stock positions.

The solution suggested in Figure 2-1 is but a reasonable extension

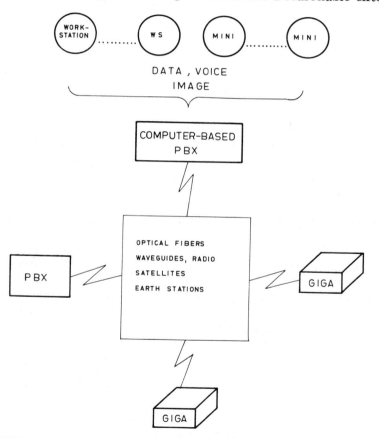

FIGURE 2-1 Workstations (WS) and minicomputers interface with the network, and through it with mainframes (gigacomputers) through PBX.

of the office environment described. Workstations and small business systems are interfaced through a PBX with the (public utility) telecommunications network. As the price of long-haul communications facilities drops, the size of communications facilities drops, the size of communications channels will widen to suit user information flows, and as more users take advantage of these facilities, the full potential of the telecommunications technology will emerge.

COMMUNICATIONS COSTS AND OPPORTUNITIES

To manage communications effectively, one must look to the near and the distant future. Telecommunications today are more manageable than ever, and managers have more useful tools available to do a better job, but future systems promise to upset totally the present values.

Which are the coming systems with a far-reaching impact on costs and services?

- Communications satellites and waveguide transmission
- Online word processors and intelligent copiers accessing databases
- Large-scale, terminal-oriented networks
- Pipelines carrying 250,000 simultaneous telephone calls
- Computer-switched telephone exchanges, which by and large, are already realities

As they grow, these systems will enable text, data, image, and voice information to be transferred a full order of magnitude more quickly, accurately, and broadly. Interactive television, allowing one to see as well as hear the other party, and two-way broadband cable (CATV) communications systems are the emerging facilities of this decade.

When perfected and accepted widely (probably in the middle to late 1980s) these developments will spark major shifts in marketing techniques, travel patterns, and work habits. Coupled to systems developments such as interactive videotex, teletex (electronic mail), and voice input recognition, they will offer strong economic incentives for executives to change basic ways of doing business.

Yet when we talk of an evolution of far-reaching consequences, we have also to ask: at what cost? Today, telecommunications is one of business's biggest and fastest-rising expenses. In the United States this cost is over $20 billion annually, with some companies spending as much as 15% of their annual sales revenues on telecommunications (although the national average is around 4%). In the past, these costs rose somewhat in line with sales, but today they rise disproportionately, which explains why most companies look upon communications expenditures as a bothersome, if necessary, cost of doing business.

That costs are rising in a particular sector is not surprising. What is surprising is that despite the magnitude of these costs, decisions in this area are often fragmented among local office or purchasing managers who have little or no technical experience in communications. The current situation can be described as follows:

- Manageable telecommunication costs are rising because of inadequate controls.

- The technology is advancing at an extremely rapid pace, opening innovative ways to employ communications to competitive advantage.

- Competitive products and services being introduced offer to every firm that has the expertise an opportunity to upgrade service levels and reduce fixed costs.

Telephone communications expenses involve both direct and indirect costs. The direct-cost elements are easier to pinpoint and analyze: the telephone company's charges for service and equipment, salaries associated with switchboard attendants, and the value of the floor space used by the telephone plant. Analyzing indirect costs, however, involves establishing general cost/benefit criteria to suit the entire organization and serve its operating departments. To examine the cost figures adequately, we must start by studying the basic service provided, the equipment supporting it, and the detailed monthly bill as well as its rationale.

Determining the degree of quality of telecommunications service is an equally important activity. This requires a comprehensive analysis of the entire telephone network, the communications load imposed by different functions within the organization, and the effectiveness of the present service. Some employees can have telephone-intensive job requirements: purchasing agents, reservation clerks, and sales personnel would actually require a higher degree of telecommunications than would the average middle manager or junior executive.

The first goal of such a study should not just be to cut costs, but to study cost effectiveness. An organization has objectives to meet: How much do the current telecommunications practices enhance those objectives? Does the system meet expectations? Is there waste in communications practices? Where? Why? How much? If the projected office automation gear is added, can the system be expanded to meet objectives as requirements develop?

We also have to pay particular attention to future costs. How much will it cost to expand or modify the system to meet the company's future needs? Can we reliably calculate the unit charges (present and future)?

Questions such as these are most important in an office automation environment because as long-distance direct-dialing experience indicates, with automatic devices operating, online costs can easily, indeed very easily, become uncontrollable. In calculating message unit charges, the following issues are important:

- *The message unit itself:* as a measure of telephone service used on station-to-station calls within an area, based on time and distance

- *The message rate service:* calculated by the computer at the telephone company switching center as message (time) unit and distance (MU/LD)

It used to be that calls within certain areas were valued at only one message unit (MU) regardless of the length of time the line was occupied. With intelligent switching centers, this practice fast disappears. Indeed, with modern packet-switching networks, the new tariffs take exactly the opposite approach. The time the line is occupied is important; distance is of no consequence. Eventually, this will have an influence on the third point:

- *Local area service:* flat rate for small area, MU/DL for others
- *Metropolitan service:* flat rate for larger area, MU/DL for others
- *Flat rate service:* more expensive flat rate for unlimited calling within basic area
- *WATS* (wide-area telephone service): a service which until recently could offer a company interesting opportunities for optimizing its telephone bills (Figure 2-2)

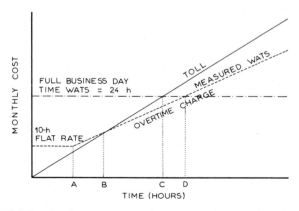

FIGURE 2-2   Break-even points in the estimate of toll and WATS charges.

The latter reference was in a way altered in 1981 as AT&T revised WATS rates and changed private line mileages. This made it imperative that effective computerized systems be employed for controlling the mushrooming growth of phone costs, starting with an accurate measure of the extent of usage. These are known as "action switches."

According to some estimates, phone costs are rising by at least 15% a year, with even greater increases to be expected as a result of the deregulation: The measurement of a call's duration is so much more important as an increasing number of telcos are persuading state regulatory agencies to tie local calls to duration instead of flat rates or a function of distance.

- *Foreign exchange (FX) service:* essentially applies one of the foregoing alternative tariff schemes from another location in the country to charge for a point-to-point long-distance line plus local charges

Tariffs vary by country and, within the same country, between ordinary and preferential rates. Furthermore, the study of tariffs and their optimization is only part of the problem. We cannot control the telecommunications budget without enforcing accountability. For a documented cost control (and billing) call records must be kept consistently and include authorization number, number called, duration of call in minutes and tenths of minutes, time of call initiation, type of long-distance line desired, and actual type of long-distance line used.

This information is the basis for billing the individual telephone network users. It also provides the data for systems usage analysis and reconfiguration. System solutions should be adopted that carefully consider the requirements. This is exactly the point where the integration of modern office gear should be seen as a replacement for, not an addition to, voice communications.

Companies that have studied the cost effectiveness of their current systems have found that they should use online text and data transmission rather than voice for recording, ordering, and information purposes; strategically located text and data terminals to help reduce clerical transaction costs; data concentrators to cut line costs; common value-added carriers rather than private lines for text and data transmission; and computer-based PBXs to achieve new combined voice, text, and data functions and to reduce operator needs.

THE TECHNICAL BACKGROUND

We will return to these issues of choice and evaluation when we talk of the challenge of choosing the right PBX and of the opportunities this challenge brings about. Indeed, we will follow a complete case study as an example of the steps that should be followed. To profit from this kind of study and help reduce the telecommunications costs of an organization while improving overall service, the manager and his or her assistants must have a good knowledge of the technical issues.

There are many subjects which may come as add-on rather than standard features, and we must know whether we need them prior to ordering them. A comprehensive terminology helps visualize the issues, but by itself, terminology will not make a telecommunications specialist. It will, however, allow one to ask the specialists the right questions and, based on their answers, judge the wisdom of one course rather than another.

Information transmitted over lines can be in the form of either analog or digital signals. Typical analog signals are voice, music, TV, and general image, including teleconferencing and picturephones. Digital signals, at least at their origin, are characteristic of telegraphy, telex, text and data communications. Facsimile can be either analog (Groups 1 and 2), or digital (Groups 3 and 4).

Whether we handle analog or digital signals, we need a carrier, basically an electrical signal chosen because of its ability to travel through the transmission medium and carry the data. (By extension we often use the word "carrier" as a synonym to the company providing this service.)

In data communications, the medium is the telephone line: a twisted wire, coaxial cable, underwater cable, optical fiber, terrestrial microwave link, or satellite. The presence or absence of a carrier refers to the presence or absence of transmitted energy.

Broadcast is the simultaneous dissemination of information to a number of stations. Recently, the term "narrowcast" has been coined to identify a short distribution list composed of intended recipients of a message.

"Bandwidth" is the difference expressed in cycles per second (hertz) between the highest and lowest frequencies of a band. Originally "wideband" was a term applied to facilities or circuits whose bandwidth is greater than that required for voice channel. The capability to modulate is critical in a wideband. In this sense, if I have to transmit only voice and telex, then the 1200-baud line is fine and any bandwidth at my disposition above that is a wideband. This is not true if the process being handled poses broader bandwidth requirements.

Television, for instance, needs 5 to 6 MHz respectively in Europe and in the United States. In this case, wideband would be at more than the 5-MHz level to allow other types of signals to travel on the same channel. (A good rule to remember is that an image calls for a bandwidth about 300 times that of voice lines, roughly 1 MHz, and TV requires five to six times as much.)

The term broadband, which we used in Chapter 1, is more specific than wideband in terms of technical characteristics. For this reason we will stick to broadband when reference is made to a well-defined process and use the term wideband in a more liberal manner. (The two terms, though not absolutely interchangeable, are treated as such in literature. Wideband is a telco term.)

When we transmit in wideband, we can multiplex on it narrower bandwidths [frequency-division multiplexing (FDM)], some dedicated to voice, others to data. Three main approaches are: DUV, data under voice; DAV, data above voice; and DAVID, data above FDM voice and video. We will return to these subjects, but let's however add that multiplexing does not necessarily need wideband to be implemented. (The CCITT rules for frequency-division multiplexing specify: for 12 channels, 60 to 108 kHz; for 60 channels, 312 to 552 kHz; for 960 channels, 60 to 4028 kHz; for 2700 channels, 312 to 12,388 kHz; and for 10,800 channels, 4 to 61 MHz.)

Another multiplexing technology is time-division multiplexing (TDM). Here the wideband is not allocated by narrower bandwidths but as a division of very small units of time. Thanks to the maturing state of the microprocessing art, a new generation of equipment has evolved, generally known as "intelligent" or "statistical TDM," which offers a significant improvement in network use. [For TDM, the CCITT rules specify for 80 channels, 2048 megabits per second(Mbps). The multiplier is >4 rather than 4 because of space and synchronization; for 120 channels, 8488 Mbps; for 480 channels, 35 Mbps; and for 2000 channels, about 140 Mbps.]

"Forward error correction" techniques are applied for detecting and reporting communication errors associated with message units. They use codes that contain sufficient redundancy to detect and correct errors without retransmission: for instance, parity or other validity checks; echo-checking is used in detecting and reporting data communications errors. The measure of errors on a line is indicated by the bit error rate (BER). For a digital transmission system, BER performance is one of the key indicators.

"Scrambling" is used to change the data so that they appear to be random: for example, to avoid long strings of "1's," which can cause modem problems, usually of the clocking circuitry. Modems equipped with loopback switches for fault isolation are used to allow the organization to determine whether a failure or an increase in errors is being caused by the modem itself or the line to which the modem is connected.

A "digital loop mode of operation" is a way of testing the modem (data set). Test equipment is connected to the near-end set and both the transmission and receiver are tested by having the receiver output connected directly to the input of the transmitter at the far end of the connection.

"Equalization" is telling the modem how to compensate for the characteristics of a particular telephone line. It is increasingly implemented through microprocessors, which allow automatic reacting to changing line conditions and minimize errors due to line distortions.

"Conditioning" refers to tolerances. A voice-quality line presents electrical characteristics conditioned by tolerances which indicate how far we can go within the specifications. Since conditioning involves decisions as to tolerances, the cost of purchasing conditioning is essentially the price paid for higher line quality.

A tremendous variety of transmission characteristics can exist on any given telephone line. Among those of immediate interest to office automation are "automatic store and forward" and "automatic dial backup." An automatic store and forward capability is used to maintain control over messages queued for a busy device such as a workstation or terminal. Automatic dial backup is important with leased lines to ensure that when leased lines fail an automatic switchover to dial facilities is accomplished for the duration of the outage.

In a workstation-to-workstation (or terminal-to-terminal) communication, "handshaking" is an exchange of signals that enables the modems to establish the communication link correctly and to synchronize operation. The handshaking sequence also adjusts the speed of the answering set.

Let us also take note of three other terms: "stream" is a virtual circuit application (in satellites) going through a single broadcast; a "window" is a logical path to be opened between two processes before data can be passed (user flow in control management); and an "incarnation number" is a unique name for an instance of the protocol mode of a workstation-to-workstation (or to a text and database) communication.

PROBLEMS IN TELECOMMUNICATIONS

What sort of telecommunications problems should a responsible executive know about? There are five to keep in mind.

The first is *transmission proper*. The choice of media, speed, and quality plays a major role in this regard. Transmission takes place on *physical* media—twisted wire, coaxial cable, optical fibers—and on *microwaves*—radio links and satellites.

When we talk of transmission, the issue of bandwidth comes up. As voice-grade lines (1200 baud) have been used for years as a unit of measurement, a distinction can be made between *low capacity* (2, 3, ..., 24 channels), *medium capacity* (up to 300 telephone channels), and *higher capacity* (from 300 to 10,000 channels and up). A better way is to talk of bandwidth in bits per second (bps).

So far we have made reference to baud and have said that it is a measure of cycles per second and characterizes analog circuits. Up to a point, the two units give the same results, but beyond 1200 baud a line may, through ingenious coding—dibits, tribits, and so on—transmit more bits per second than the baud it supports.

With digital circuits, bps (bits per second) is the measure and, in terms of bps for advanced telecommunications usage, we now speak of normal capacity up to 56 kbps (thousand bits per second), with 100 to 400 mbps (10^6 X bps) being the high capacity under present conditions. The latter is characteristic of satellite transmission.

The second problem in telecommunications is *distribution.* Signals have a destination; they must be channeled sender to receiver(s). The mission of a telephone exchange, for instance, is to answer the distribution responsibility.

Distribution can be *two-way* (bidirectional) or *one-way.* The needs of two-way distribution brought about the development of switching mechanisms (step by step, crossbar, electronic switches). Two-way solutions are necessary for interactivity (conversation).

Some processes do not require bidirectional switching, as is the case with TV and radio (music) broadcasts, the telegraph, and to some extent, facsimile—particularly for group 1 slow speeds. The one-way solutions (such as telegraphy) were the earliest to be used.

In general, present-day networks are *not* able to homogeneously transmit and distribute the various types of information with which we deal. With the evolution of technology came the third problem: *multiplexing,* the regrouping of signals in one-way transmission.

Physical media, (coaxial cable, optical fiber) permit a high level of regrouping. The same is true of satellite broadcasting. The practical limit is set by the interferences which exist. In the preceding section we referred to FDM and TDM as the best examples of multiplexing technologies.

The fourth problem to keep in mind is *electrical characteristics:* conditioning, synchronization, regeneration, and the handling of the frequency spectrum. As we progress through the electromagnetic spectrum, even the nomenclature changes:

- "Discrete frequencies," mostly in audio and speech.
- "Conventional wavelengths" refer to microwaves, a more recent trend being toward millimeter waves.
- "Rays," which can be heat, light, or molecular-electron emissions, are also frequency-related phenomena.

We spoke of wideband regarding media that allow the transmission of any type of signal, but wideband also presents technical problems—if not in transmission, at least in switching. Present-day technology has limitations. In the case of switching, the switching centers (telephone companies) are *not* made for transmitting TV signals; they are made for voice and therefore work at about 4 kHz, whereas TV needs 5 MHz.

The fifth problem in telecommunications involves *protocols*, formal sets of conventions governing the format and the control of data. Protocols constitute logical levels of connection to the physical line (carrier). Simple early protocols were of the "start/stop" variety. Modern protocols include advanced concepts. Primarily, they standardize known, *not* new things. As such, they have two functions:

1. *Contact:* Identification, synchronization, and creation of a virtual (logical) channel. These facilities concern source and destination.

2. *Transfer:* Including EDC (error detection and correction), store and forward, and the assurance of delivering the message.

Networks of links, processors, and databases need a protocol with which to work. It is also necessary to provide gateways for equipment following different line disciplines so that they may communicate among themselves. For example, most organizations will have a line that has start/stop and bisynchronous terminals on it and also want to use packet-switching units, or plug that original transmission line into a larger network that is basically packet-oriented.

Let's conclude this discussion by underlining the importance of protocols in the future of computers and communications. Neither hardware nor software is central to the study of man-information communication channels, though both are certainly necessary for implementation. The most fundamental component is the *communications protocol.*

Chapter 3
IMPLEMENTING AND USING THE NETWORK

INTRODUCTION

The major elements in the telephone network are:

1. *Station equipment*
2. *Transmission facilities:* loop, exchange area, long haul
3. *Switching facilities:* local and toll

Data communications systems designers have long anticipated resources now available. The information flow, expressed in bits per second, can approach an upper limit known as the "channel capacity," provided that the proper encoding is used. Encoding also sees to it that the channel can function with a vanishing probability of output error, despite the inevitable noise disturbances present in all communications systems.

The search for channel capacity also led to significant developments in switching technology. The switching process necessitates activation of some sort of physical connection or path to allow a conversation between two stations (telephone sets, workstations, terminals). Switching can also perform alternate routing. It can be computer-based and offer facilities such as store and forward.

Six functions are implicit in the basic connecting capability: alerting, attending, information receiving, information transmitting, busy

TELEPHONY NEEDS:

SUPERVISORY SIGNALS   NETWORK MANAGEMENT
ADDRESSING   ERROR CONTROL
INFORMATION

THE   PACKET   STRUCTURE   IS:

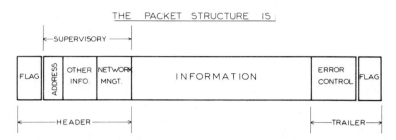

FIGURE 3-1   Projecting an approximate relation between the prerequisite of voice communications and the packet structure.

testing, and supervising. Some involve activities both at the station and at the switching level, and there is a correspondence between telephony's prerequisite and support services and the structure of a packet-switching network (Figure 3-1).

The interchange of information is known as "signaling." Signaling overlaps with switching because the elements used to send and receive signals are sometimes considered part of the switching system.

Signals are classified into supervisory, address, information, and network management. The last is a highly structured proposition and, as we will see, involves distinct hierarchy levels.

LEVELS OF HIERARCHY

The public telephone network has five levels of hierarchy in switching offices (Figure 3-2), seven if station equipment (PBX and sets) is included. As far as the public level is concerned, "loops" (the transmission paths between station equipment, or PBX, and the serving central office) are the lowest level.

"Exchange area systems" are used to provide short trunks. They differ from long-haul, being designed for less stringent performance objectives and thus achieving economies.

"Long-haul systems" are high-capacity carriers, providing long circuits. They can be coaxial cable, submarine cable, microwave links, satellites, waveguides, and so on.

Switching facilities are divided into two categories: local and toll. Local facilities are those at the central office to which station equipment is connected directly by loops. Toll switching systems connect

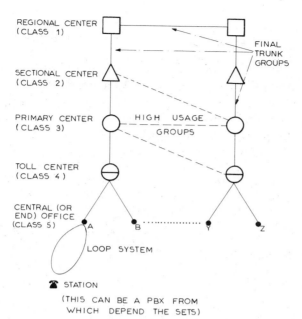

FIGURE 3-2   Levels of hierarchy in a classical voice-grade telephone network. Each switching office is connected to an office of higher level.

trunk to trunk. "Tandem" is a special class of trunk used in large metropolitan areas to interconnect local offices.

Generally speaking, the total investment in telephone facilities can be broken down into four classes:

- Station equipment: 20%
- Switching facilities: 23%
- Transmission facilities: 45%
- Land, building, and equipment: 12%

Besides the public service lines, telephone companies have leased lines, and value-added networks have rented shared services. Private line offerings include: telegraph, teletypewriter, remote metering, voice, data, voice and data, telephoto, audio (at 15 kHz bandwidth), television, and general wideband facilities.

An example of a private network is "autovon," a military worldwide switched network for voice and data. An interesting feature of autovon is "preemption," which allows a higher-precedent call to preempt a trunk or line.

Station equipment includes investment in sets, key sets, PBXs, word processors, teletypewriters, terminals, computers of all types, mobile radios, and modems. About one-fourth of the investment in switching facilities is for toll systems, and one-third of transmission facilities' investment is for long-haul.

The sharing of such expensive devices has been a steady preoccupation of communications engineers. As a result, combined voice and data communications networks are a growing trend in today's information systems. But aggregate systems work best if we can statistically analyze and project the operational behavior of a known traffic pattern on a communications network.

Statistical algorithms address these requirements, projecting such communications traffic behavior. Assuming that there is a random voice and data communications calling pattern, the combined pattern is segmented into a representative period (such as the average usage or the peak usage hour).

On the control and optimization side, computers provide automatic route selection and queuing capabilities if the facilities are busy at the time of selection. The major concern is then the definition of the proper type and quantity of network facilities to meet economic and service objectives.

The paramount unknown to be determined is the probability of "blocking"; that is, an event finds all available paths or routes in use. Where such blockage occurs, the automatic building and distributing of a queue must also be determined and the amount of traffic must be statistically definable. The difference between *queued* volume and *recovered* volume is *queued overflow*.

THE HANDLING OF TRAFFIC

The handling of traffic can be viewed in terms of *attempts* and *holding time*, and call has two cost components:

- *Establishing the call*, which involves complex equipment over short time periods
- *Holding the call*, which requires relatively simple switches and logic circuits, but for a longer time

The structure of the facilities is governed by a number of factors: the available communications technologies, the location of the customer, performance objectives, the need for redundancy and reliability, and the cost and availability of land. At any given time, however, the use of the available facilities must be optimized, and for this we need a unit of measure and a database where the results of measurements can be stored and retrieved.

The unit of measure must focus on the *intensity of traffic*. CCITT recommends the erlang as such a traffic unit. The erlang measures the number of circuits contemporaneously occupied in an instant t and provides a basis for calculating use availability based on the actual telephone (voice, text, data) traffic statistics.

The number of erlangs that passes in 1 hour in the circuit represents the *percentage of time* the circuit remains occupied. This is expressed algorithmically as follows:

$$Nt_m = A \quad \text{(erlangs)}$$

where N is the total number of communications during the peak hour and t_m is the average time of each.

For instance, 0.3 erlang means that this circuit is occupied 30% of the time. Thus this unit of measure expresses the *medium intensity of traffic*. The product is a function of a cascade which indicates successive accessibility. The larger the accessibility, the easier it is to distribute the traffic.

Demand in a selected hour is the sum of all call durations: the total time required to establish and complete all selected hour calls. The American unit for expressing demand is "hundred call seconds" (CCS). There are 3600 seconds, or 36 CCS, in 1 hour. Whichever of the two units is used, the fact remains that communications principles, applied to a switched telephone system, have resulted in a calculated grade of system service based on an average or peak period of operation. Two demand characteristics must be defined:

1. The duration spectrum of the event or events that would be expected to occur during the selected time period

2. The relation of the selected period of time to total demand

In network system design, the first is usually the total amount of required time represented by the "demand" during the selected period of elapsed time. The second characteristic is, typically, the representative hour of operation and can therefore be regarded as the application of a worst-case design philosophy.

Logically, if the system can accommodate the peak period with a satisfactory grade of service, all other periods of operation will enjoy an even better grade of service. It should be remembered, however, that this relation takes into account neither economics nor other realities, and the term "satisfactory" is not always subject to a purely objective definition.

TRADE-OFFS

The communications technologies evolved rather slowly during the first 100 years of telephony, but the pace has accelerated tremendously during the last decade, both in switching and in transmission. Merging different technologies, and also different types of services, calls for interfaces. An interface is a common boundary where two systems, or pieces of equipment, are joined. A basic function of an interface is to provide a set of points that will, to some degree:

- Separate responsibilities on the two sides.

- Allow each side the flexibility of rearrangements and the evolutionary introduction of equipment and services.

The evolutionary introduction of new technologies allows us to capitalize on them without upsetting the whole system. Furthermore, there is a degree of interchangeability between switching and transmission. The most common instance of trade-off is concentration.

Switching is used to reduce the number of transmission channels necessary to provide a given service. Alternatively, when there is enough traffic between a pair of points, a private line without switching can be the best traffic network. FX (foreign exchange) is an example of using transmission to save switching.

Storage capability (a store and forward system) can be used to save transmission. Storage is usually added to the switching equipment, but we can also have intelligent lines.

These are the basic premises on which private line engineering capitalizes to optimize available (and projected) facilities as a function of the number of calls to be handled, the actual or expected average call duration, and the circuits (or trunks) and switching centers.

Criteria to be used in studying a network are directly dependent on the average call duration and the total demand for available circuits. In the event of an all-circuits-busy condition, the call must be queued until a circuit is free.

Delay on queued calls is determined primarily by:

1. The average call length
2. The number of possible paths that could be available to process the waiting call

Once the selected-hour demand has been established, it is assumed that this demand will be processed without any blockage, which could only be introduced by the public telephone network. In such an event, the call would be reintroduced automatically. Communication common carriers define network service in the context of these terms.

Grade of service is conditioned by the probability of finding all trunks busy, and this probability is the basis of the circuit capacity tables, which identify grades of service as P01, P02, P03, and so on, to indicate the probability of a call finding a circuit busy during the selected hour. For example, P01 signifies a probability of 1 in 100 calls of finding a busy condition, P02 signifies 2 in 100, and so on. With the P10 grade of service, 90 calls in 100 will find idle trunks on the first attempt. A good grade of service for the public telephone network is P02, using Poisson distribution. (Specifically, Poisson probability is used where the number of trunks or facilities provided assures a minimal percentage of blocked calls.)

The selected-hour load to a group of circuits is the amount of traffic (demand), expressed in terms of CCS, which will receive the optimum grade of service. The extent to which total available circuit time in the group is actually used during the hours selected indicates the group's efficiency and loading.

DESIGN CHOICES

When it comes to networks for voice, text, data, and image, we need rather sophisticated means of making a company's various machines talk to one another. As most companies have quite chaotic communication networks, built up piecemeal to fit particular purposes and lacking coherent design, Figure 3-3 suggests a decision matrix that has proved to be helpful in evaluating the state of the art and the principal tendencies in projecting a network structure.

		ONE ORIGIN ONE ADDRESS	ONE ORIGIN MANY ADDRESSES	MANY ORIGINS ONE ADDRESS
AGE OF NETWORK	OLDER	✓		
	NEWER		✓	
PRINCIPAL CONTENT	DATA	INQUIRY ✓	BROADCASTING ✓	DATA COLLECTION
	TEXT	ELECTRONIC MAIL ✓	CIRCULAR LETTERS ✓	REPORTS ✓
ORIENTATION	CLASSICAL VOICE	✓	TELECONFERENCING ✓	✓
	VOICE MAIL	✓	✓	
ANSWER TO MANAGEMENT NEEDS	ORDER GIVING	✓	✓	
	REPORTING	✓		✓
PUBLIC USE	BROADCASTING VIDEOTEX		✓	
	INTERACTIVE VIDEOTEX	✓		✓

FIGURE 3-3 A decision matrix for evaluating the main tendencies in projecting a network structure.

Our objective should be to search for ways to unify network components and gain much easier access to text and databases. To profile accurately the expected range of possible network performance, we must make calculations using a maximized call length and traffic distribution by peak day and peak hour. Conversely, a projection on average usage must be made through the same parameters. The results indicate expected network behavior. Through an iterative computer program, it is relatively simple to develop expected network performance envelopes with a range of parameters to provide the basis for cost-effective design.

Designing suitable communications networks that consider all needed factors is relatively complex. Quantitative models have been developed to determine the minimum-cost network under given traffic patterns; but although these models are very useful, they are not sufficient. For optimal design we must consider some important intangible factors that may strongly influence the choice of communications alternatives. In doing so, we must always recall that the complexities of text and data transmission are too great to warrant a single general answer over a large variety of applications. Furthermore, the parameters of choice associated with a particular requirement must be examined to decide whether solutions appropriate to a given need can be met by an available public offering, or if the requirements can be changed to match those of a public offering.

Among the outstanding problems relating to the establishment of communications networks for text, data, image, and voice using the latest technologies is the question of how well the system we project is going to integrate with the existing telephone company (telco) voice-grade networks. Is it rational to put horizontal network (user) solutions on basically star telco networks? Table 3-1 presents a list of advantages of packet switching versus message switching. Figure 3-4 compares the use of three distinct technologies: space-, frequency-, and time-division

TABLE 3-1   Advantages of packet versus message switching.

	Packet	*Message*
Speed of the lines	Fast	Subvoice
Quality of the lines	High quality	Aging
Node intelligence	Computer supported	Early store and forward capability
Packet/message size	Standardized (data container)	As is (telex varies from 20 to 1000 channels per minute, with a 66-channel mean)

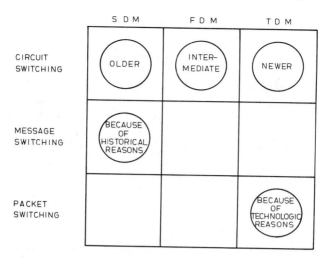

FIGURE 3-4   The use of three distinct technologies in networks.

multiplexing (SDM, FDM, TDM) in networks supporting circuit, message, and packet switching.

Circuit switching is the classical approach used in telephony for connecting the point of origin to the point of destination through the assignment of a physical line which remains dedicated as long as the telephone connection lasts.

Message switching was advanced after World War II with telex-type applications. The modern demand for telex lines and a wider variety of services, such as call transfers and broadcast calls, are new values added to the message system in response to the accelerating speed and diverse ways in which information must be exchanged on the domestic and international levels.

Packet-switching systems will provide the answer in the near future. They provide an efficient communications means through logical circuits able to optimize the use of resources and to improve reliability. The stimulus for this is the need for people in faraway places to stay in touch, which, in turn, means an ever-wider variety of international communications services. A packet-switching system represents today's most advanced integration of telecommunications and computer technologies. It can handle online computer data, facsimile, and telex/telegraph, adding a new dimension to worldwide communications exchange.

Other questions to be asked are: How radically will the current network concepts be altered by the coming "user" requirements of 10,000, 20,000, or even more terminals and workstations? How can we protect our investments while keeping the door open to the steady stream of technological advancements?

An economically pertinent question to ask is whether the volume of traffic involved is sufficient to stimulate the construction or lease of a special-purpose network. It is important to consider the answer to the question: What are our alternatives?

If a value-added network can offer the services we want, and they correspond to our topology, we had better use these services. If not, we shall have to build them ourselves or provide links to the available nodes.

Should we decide to build a COAM (company operated and maintained network), the basic building blocks (backbone) of the network will be nodes and lines, say, 56-kbps digital or analog trunks with tandem-switching traffic handling. The simplest solutions are the best, and this means a clear separation of:

1. The transport facility
2. The text and data processing facility
3. The text and database

The nodes must be able to execute system programs and provide facilities for storage, EDC, recovery, and so on. The network should provide a "framework" within which users can tailor subnetworks of computers and terminals to fit their specific applications. It must also permit easy configuration of subnetworks as requirements develop.

Network software must ease the burden of communications (transport) management by providing information on subnetworks status, usage, and performance. It must be adaptable to the logical arrangement of workstations, terminals, and host computers. And it must support a variety of station classes reflecting a broad range of characteristics, including:

- Slow speed through high speed
- Synchronous and start/stop
- Character and packet mode
- Clustered and standalone
- Commonly used code sets (ASCII and EBCDIC)
- Polled and contention line control

The network must provide gateways. Current and projected protocols in use by the organization should be supported. It must be possible to connect hosts and terminals via point-to-point or multipoint private lines from remote premises by means of either digital or analog facilities, and remote workstations and terminals must also be connected using dial-up access through the public network.

Chapter 4

SWITCHING AND TRANSMISSION

INTRODUCTION

The telephone exploits a basic electroacoustic technology that is difficult to improve on, given very tight economic bounds. Simple as this terminal is, it acquires enormous power by virtue of the network to which it provides access. That network embraces more than 200 countries and territories, making it possible for virtually any pair of telephones among the more than 350 million to be interconnected on command. The new needs (data solutions) extend this perspective.

A basic reason for so much interest in exploiting to the fullest degree the potential of a telephone system is its general use. Since it has been domesticated, the telephone reaches many strata of a working population. Many unexploited areas also exist in telephony. For example, a simple attachment to the telephone handset makes it possible to transmit a patient's heart signals to a physician's office, where the signals are produced on a cardiograph.

But as we stated in the first three chapters, in its most fundamental structure this worldwide telephone system uses three basic components: station equipment, transmission, and switching. Understanding how these subsystems function is almost synonymous with getting a comprehensive view of how telephony works. The increasing concern for effective and efficient teleprocessing support obliges us to look toward the future evolution of this system.

Text usage of computers and communications networks may swamp data usage. The two together, however, constitute complementary parts of document distribution. Voice communications as we know them today will probably taper off, but voice store and forward, teleconferencing, and voice-over-text and data capabilities may have an exponential growth. All of this adds to an increasing use of telephony and makes mandatory an understanding of the mechanics of the system.

THE NUMBERING PLAN

Dial telephone service has many prerequisites, but none more fundamental than a *numbering plan*. This numbering plan establishes the structure within which dialing procedures can be applied. It must accommodate:

- All authorized address devices
- All users, including those concerned with system testing and maintenance

Dialing must follow a prescribed sequence. First, there may be a need for one or more prefix digits; next, the address must be transmitted; and finally, a suffix may apply. The basic address format (used in North America) consists of 10 digits with a 3-3-4 subdivision, corresponding to (1) the numbering plan area code, (2) the central office code, and (3) the station number.

With international direct distance dialing (IDDD), standards become a matter of global concern. In 1964, a CCITT study established 11 digits as a preferred maximum length for international numbers. The following world zone assignments have been made:

1	Canada, United States
2	Africa
3, 4	Europe
5	Central and South America
6	South Pacific
7	Soviet Union
8	North Pacific
9	Far and Middle East
0	spare

But there is no standardization in the way the dialing is done for international connections. Germany, Switzerland, and Italy use "00" to access an international line and "0" to access a national long-distance (area code) number. Correspondingly, France uses "19" and "16," followed by the international or area code. Other countries have yet different standards. It is indeed regrettable that the PPTs (the European postal, telephone, and telegraph authorities) and telephone companies have not gotten together to standardize issues that should never have presented so many permutations and variations in the first place.

TRANSMISSION MEDIA

A "channel" is a transmission path dedicated to providing communication between two points. Over the channel facilities can travel analog information, video signals, or digital data.

Classically, voice has been transmitted in an analog fashion, but the new technologies promote digital voice transmission. A signal time function is sampled at regular intervals and at a rate at least twice the highest significant signal frequency (8000 times per second). The samples contain all information of the original message. Reducing a signal to a limited number of discrete amplitudes is called "quantizing."

There are basically two types of video signals carried over telephone facilities: television and picturephone. As we stated, television requires a bandwidth of about 5 MHz (more precisely 4.3 to 5 and 6 MHz), the picturephone a bandwidth of 1.0 MHz, and we said already that an easy way to remember this is that it is a band nearly 300 times broader than voice.

Text and data networks can use either analog or digital facilities. With analog facilities, processing is required to permit the transmission of digital signals. This is called "modulation," and it is used to enable digital signals to be transmitted over analog channels that would destroy the digital waveform.

A sort of modulation equipment (vocoder) is used to convert voice or voice-like signals to digital form, which enables lower noise or distortion and economies in system implementation. The use of digital facilities started in 1974.

With "Dataphone Digital Service" (DDS), text and data from a word processor, terminal, or mini, micro, or personal computer are transmitted end to end using digital techniques. This reduces complexity and cost, and it increases transmission quality, throughput, and reliability. As a matter of fact, the object of DDS is to provide a communications medium for transferring data between computers, workstations, and terminals. The signal is not amplified, but regenerated. This ensures the data integrity in the original data form.

Automatic performance monitoring is incorporated into critical components of the system. Testing and maintenance features are built in to permit a network control center to test and isolate trouble on both ends of a digital channel. A specially designed data service unit performs proper coding and decoding of signals, formatting, timing recovery, and synchronous sampling.

Using DDS is simple. The 2400-, 4800-, and 9600-bps models have an interface that conforms with the RS-232-C standard. The 56,000-bps model conforms with CCITT Recommendation V.35, and its control signals conform with Standard RS-232-C. All models should therefore be directly compatible with any business machines that conform to these standards. Current speeds are 2.4, 4.8, 9.6, and 56 kbps, and 1.5 Mbps. With current technologies, satellite transmission will permit 100- and 400-Mbps widebands.

(It is proper to add at this point that when it comes to "standards" there exist a number of them: old and new; American, European, and other. CCITT has adopted the V35 standard at the level of physical interconnection. Its electrical characteristics are: ±0.55 V on 100 Ω and it supports 34 pins. It is double ended, data only; and it is mainly employed in Europe.)

(The older American recommended standard is the forementioned RS 232, supporting 25 pins in "D" shell connector. It is single ended, data only. The newer recommended standard is RS 449 and it supports 37 pins. The RS 449 is double ended—data and control—and divides into two forms. The RS 422 is balanced; the RS 423 is unbalanced.)

Digital data signals involve machine–machine or human–machine communication. In one form of transmission the pulses may occur only at regularly specified times: *synchronous*. An alternative form is *asynchronous* transmission, which puts no restriction on the pulse length or time of transmission.

Synchronous and asynchronous are line disciplines necessary for the orderly use of the resources available for transmission, that is, the *transmission media*. Data transmission in itself varies from a few pulses per second (for supervisory control channels) to over 1 million pulses per second.

There are seven principal types of transmission media:

- *Open wire lines* consist of pairs of uninsulated wire strung on poles.

- *Paired cable* is the well-known twisted pair protected by plastic, aluminum, lead, and so on.

- *Coaxial cable* consists of an inner copper wire conductor insulated from a cylindrical outer conductor and is now widely used.

- *Waveguides* provide a transmission system of vast potential because, in operating over a circular waveguide, a very wide bandwidth can be achieved.

- *Optical fibers* comprise one of the most promising technologies. Through lasers, we have found the way to harness light waves.

- *Satellite radio* is, because of its excellence, the sideband medium of the future, able to provide a wide range of domestic and business lines and message services.

- *Terrestrial radio* is also a leading transmission path. The main advantage of radio over wire is the absence of a need for physical facilities between the two points of communications.

With radio, the available received power is limited by the transmitter power, antenna patterns, path lengths, atmospheric conditions, and other obstacles. Solutions have always been found to overcome these problems. Radio systems used or planned operate at 4, 6, 11, or 18 GHz.

All told, satellite radio, optical fibers, and waveguides are the newest transmission media with the best future. They present significant advantages in cost effectiveness—broadband capabilities and sharply reduced cost figures—and promise to revolutionize the transmission side of telephony. A recent study indicated that when the new generation of communications satellites is in orbit, transmission costs may drop to 2.5% of their current figures.

SWITCHING GENERATIONS

The whole concept of dial or automatic switching is the basis of telephone communications today. When one of the millions of telephone users lifts a handset, switching equipment located in the central offices throughout the country:

- Locate and identify the calling line.
- Give the signal to proceed (dial tone).
- Determine how and where to get access to it.
- Locate and test the numerous transmission paths leading to it.
- Select and link up the most appropriate combination of these paths.
- Then, if it is not in use, ring the called telephone.

A serving vehicle has carried out the switching function. Since 1891, when the Strowger automatic system was publicized as the "girl-less, wait-less telephone," switching has taken big steps, most of them during the last 20 years. We distinguish three main switching generations:

1. Step by step, direct, or progressive
2. Crossbar
3. Common control

The last divides into electronic and computer-based solutions.

The step-by-step system was born before the beginning of this century. It was the basis of the first private branch exchange (PBX). With the introduction of the step-by-step PBX, "number, please" manual switchboards rapidly disappeared. Intraoffice conversations were established by direct dialing.

The first-generation switching systems were also called POTS (plain old telephone systems), and many businesses requiring dial switching still use them. Such systems can be expanded indefinitely as long as space can be provided for the bulky frames and switches they require. Through careful maintenance they still provide economical (minimum cost) and dependable telephone service, although they offer limited features.

In step-by-step switching, a call progresses one step at a time as the telephone user dials each successive digit of the number. The system is also called direct control because each switching function is directly controlled by the pulses from the dialing telephone. The switch train is composed of:

- The line finder
- The selector
- The connector

The dial pulses from the calling telephone directly control the switches that establish the desired connection. This is a simple, economical, and completely modular solution, but technologically speaking, it is obsolete. Its disadvantages are high labor content, switching delays, no economies of scale, and a lesser suitability for text and data communications (because of stability and noise problems).

Furthermore, step-by-step equipment is bulky and requires a high degree of preventive maintenance. Its design imposes significant limitations on the flexibility and capabilities of the telephone system it supports. Step-by-step offices are usually noisy, causing more impulse noise on subscriber lines than do other switching systems. (Impulse noise, while not significant to voice, translates into errors in text and data communications.)

Corrections of possible errors by error control techniques reduce throughput, that is, the amount of data that can be transmitted in a given amount of time. Other deficiencies of step-by-step switching equipment are the time required to complete a call, the tying up of all switching equipment associated with a call until completion of the call, and the inability to go back to the previous switching stage to look for alternative connecting paths in case of a busy condition.

Common control employs logic circuitry. Address digits generated by the dialing instrument are stored, translated, and flexibly used for switching within the system or for establishing a connection with the outside network. The switching equipment stores the entire number, then the operation starts.

"Crossbar" has been a milestone preceding the electronic common control technology, and many consider it as the first phase of common control. The crossbar switch is much smaller than the step switch. When they were introduced, crossbar systems provided many helpful new features, but they have been overtaken by electronics. Five key components make up this system:

- The *marker* is the portion of the switch through which all calls must pass. It identifies a line-requesting service and assigns to it an originating register.

- The *originating registers* record the number dialed.

- The *register/scanner* is a dual circuit in electronic crossbar systems which may perform the marker and originating register functions.

- The *matrix* consists of a set of horizontal and vertical bars. To close a set of crosspoints, the horizontal bar moves first. The point at which these two meet establishes the connection.

- The *sender*, or *trunk interface unit*, is the equipment used to process calls from the PBX to the serving central office.

Common control equipment makes it possible to adopt flexible numbering plans to meet a carrier's requirements and, in the case of private interconnect systems, to meet many specific user applications.

Common control equipment also permits new features to be readily added; they need only be applied to the circuits. Such equipment is relatively easy to maintain, requires less space than that used in the step-by-step system, and is rather effective in its usage.

Even when we talk of crossbar technology, at the central office switching level the sender transmits the called number to the different types of distant, central office equipment. All told, we speak of a specialization of functions which largely revamped the way switching was working with the step-by-step system. As in the 1970s, however, rapid technological advances in computer design channeled into telephone switching have changed the system further.

ELECTRONIC SWITCHING SYSTEMS

Electronic switching systems (ESSs) offer the greatest potential for both voice and data communications, together with the capability for an almost bewildering array of internal service features. An ESS consists of:

- A computer
- Memory or storage
- Programming capability
- An extremely rapid switching component

The principal advantages of the stored program are that it allows the system to expand its capability to perform self-diagnostic checking and automatic reporting of malfunctions, enables the user's technicians to perform many system changes, and permits inputting new requirements through teletypewriter terminals rather than by manually rewiring various switch-point connections. This leads to significant flexibility for telephone company subscribers and private interconnect users, while smaller physical space is required to house the basic switching apparatus.

A computer-based common control switching equipment implies two distinct types of units:

- Control
- Switching

The common *control* receives, stores, and interprets dial pulses, and then selects an available path through the *switching* hardware to complete a connection. Parallel processing on common control equipment for only a portion of a call is very important. Once the connection for a call between two telephones has been made, common control releases and can complete more calls while the other subscribers are talking.

Efficient, high-speed common control equipment can complete many calling connections during the time of an average conversation. Thus economies of scale are possible. This is a prime difference between common and progressive (step-by-step) control switching systems.

Furthermore, the switching network can be directed for many lines by one common group of control devices. The control unit is the brain of this switching system; it can typically complete its function for a single call in a small fraction of a second, allowing it to service many stations and lines.

An important element of the various ESS offerings is the computer's ability to perform a wide variety of traffic analysis and telephone-related accounting functions. For these very reasons, electronic switching systems have found an expanding market in PBX at different levels of sophistication.

ESS is not the latest in developments, and electronic switching is still performed by electromechanical devices. Improvements come one at a time. First, wire relays provide the logic required for supervision and control over the system. Sometime later, computer-based software is used for control and supervison.

Essential to the performance of switching functions through solid-state circuitry is that no moving parts are employed, but the basic concepts do not necessarily change. Many of the features incorporated into electronic switching originated at the crossbar era. Let's have a look at them.

Station transfer sees to it that the user can transfer an incoming call from outside the office to any telephone within the crossbar system. This eliminates the use of a switchboard operator for call transfer. *Consult and hold* assures that an incoming call can be held while the person dials another number to secure information for the caller.

Through an *add-on conference* a third person may be dialed so that the outside caller, the call recipient, and the third person may conduct a three-way conversation.

Camp-on is an interesting feature. If the user's line is busy, the operator can "camp" an incoming call onto the station line. When the user hangs up, that telephone rings and the waiting call is connected. When the call is camped on, the user hears a "beep" indicating that a second call is waiting.

Night answer from any station provides that after the switchboard closes, audible signals announce calls arriving on central office trunks. Any person working after hours can dial a code, pick up the incoming call, and transfer it to the person sought. With *selective toll restriction*, a particular station is permitted to make local calls, but the equipment rejects any long-distance dialing.

All crossbar systems offer these features; however, for lack of training, a large percentage of employees are not aware of them. The employees revert to traditional call transfer procedures involving the operators. Once again, continuous reeducation of telephone system users is important if expensive features are to be worthwhile.

More sophisticated capabilities are possible with switching centers that are computer-run. The computer is programmed to produce almost any feature designed by the user. Such facilities are contained in the central switch, not in the telephone instrument. The user commands the computer by dialing specified codes related to each available feature. In the more expensive electronic telephone systems a function key is pressed to energize a desired attribute.

INTERCONNECT

The need to provide interconnect capabilities is not at all new to telephone companies. Approximately 1800 independent telcos are interconnected in the limits of the major telephone utilities in the United States. Many of these, especially those serving predominantly small, rural areas, are relatively minor operations in which total subscribers probably do not equal those of a typical large corporate facility.

On the other hand, a number of manufacturers in the United States and abroad supply telephone equipment to the market to satisfy user requirements. Such manufacturers, however, were prohibited from attaching their equipment to the public switched network operated by common carriers. A section of the telephone tariff on the so-called "foreign attachments" states:

> No equipment, apparatus, circuit or device not furnished by the telephone company shall be attached to or connected with the facilities furnished by the telephone company, whether physically, by induction, or otherwise. . . .

In the last 15 years this situation has changed; after the Carterphone decision, change began to snowball.

First, in 1956, came the "hush-a-phone" case. The decision by the U.S. Court of Appeals in favor of hush-a-phone is the earliest court decision in the field of interconnect. The hush-a-phone was cup-like and non-electrically connected. Functionally, it was attached to the mouthpiece of the telephone set. Its manufacturer claimed that the device provided privacy to users by making their voices less audible to nearby persons.

In its decision, the court specifically noted that the "foreign attachment" provision of the tariff was "an unwarranted interference with the telephone subscriber's right to use his telephone in ways which are privately beneficial without being publicly detrimental." As a result, the court ordered AT&T to revise its tariffs, which it did.

The new tariff still retained a general provision against foreign attachments and specifically banned devices that involved direct electrical connection to its facilities. The Carterphone case came 12 years later, in 1968.

The Carterphone was an acoustically coupled device used for "patching" conversations between mobile radios and the dial-up telephone network through a base station. The interface between the Carterphone and the telephone lines involved acoustical coupling.

In 1966, Carter Electronics filed an antitrust suit against the Bell System and several independent phone companies, and the case was brought before the Federal Communications Commission (FCC). The legal battle dragged on for two years until the FCC, in July 1968, ruled against the telephone companies: "The tariff has been unreasonable, discrimination and unlawful in the past, and the provisions prohibiting the use of customer-provided interconnecting devices should accordingly be stricken." The commission went on to say: "A customer desiring to use an interconnecting device to improve the utility to him of both the telephone system and a private radio system should be able to do so, so long as the interconnection does not adversely affect the telephone company's operation or the telephone system's utility for others."

The third test involved *Magicall* and *Phonetele*. Data Corporation manufactured a call diverter device and the Magicall equipment, which it sold to the telephone companies, which, in turn, installed and leased the equipment to subscribers.

In such applications, no interconnecting arrangement is required. Users obtaining their own Magicall equipment, however, must have their units connected to the dial-up network through a carrier-furnished VCA/NPD and pay a monthly charge.

In April 1973, the California Public Utility Commission (CPUC) ruled that General Telephone Co. of California must permit the interconnection of the divert-a-call device directly to the dial-up network, subject to assurance that the device does not interfere with the telco's network integrity. The CPUC found that the requirement of VCA/NPD for use with user-owned divert-a-call amounted to a form of trade restraint and that the coupler was not necessary for network protection.

These three now classical decisions have opened the new frontiers of present-day interconnect. They have provided the grounds for a golden horde of competitive companies and products which, since Carterphone, have entered the telephone equipment market.

Chapter 5

THE PRIVATE BRANCH EXCHANGE

INTRODUCTION

Communications, as we said in the introduction, is a fundamental characteristic of our society. The integration of telecommunications (voice, text, data, image) with word processing, transmitting, copying, and data processing created the broad and potent discipline of telematics, but we cannot master this great field, or make any profit from it, if we do not have a comprehensive understanding of telephone communications.

We have made it clear that telephone communications are composed of three key parts: the station (or terminal), the transmission facilities, and switching. There is a public network covering each country, each continent, and the whole globe. There are also many private networks which, typically, connect the terminals (voice, text, data, image) of a factory, an office building, an apartment complex, or a geographic area. Private users at one end (a station) can communicate with other private users by using public or private facilities.

When we talk of an office building, a factory, and, in general, company premises, we need an interface to connect to the public (or private) carrier. This interface, the PBX, can have behind it a significant variety of terminals: voice, text, data, or image. That is why the computer-based PBX has become a pivot point for office automation.

Exchanges of this sort exist today, but their capacity will need to be notably extended to include all computer-supported functions needed in an office environment. The PBX of the 1980s will be the brain of a multiplicity of functions, which it will assure permanently. One of the functions is voice store and forward; another is data concentration (both multiplexing and local intelligent capabilities); a third is image handling from facsimile to video presentation.

The telematic PBX will sort the information coming from terminals (whether text or data), memorize it, provide for local processing or transmission on a real-time basis, and generally interface the users and their local computer resources toward the central computer and telecommunications facilities.

PRIVATE BRANCH EXCHANGE
FOR OFFICE AUTOMATION

The office switchboard market starts looking like the hotly competitive world of computers. Dramatically changing technology has virtually made America's 220,000 private branch exchanges obsolete. The recent PBX generations are computer-based.

Over the last 10 years, both the oldtimers (telcos) and the newcomers in this market have introduced fully computerized models, supported by a fair amount of software. The newcomers, the "interconnect companies," hope to boost their share of PBX installations from 6% in 1976, to 15% in 1982, and to more than 20% in 1985. The oldtimers fight to hold their reign on a market that is so fast growing.

Estimates of the size of the annual market for new installations and replacement PBXs stand at more than $500 million in 1980, with an annual growth rate better than 15%; however, by 1982–1983, with the satellite business well under way, this figure may radically change.

Current estimates for satellite earth stations range from 40,000 to over 400,000 units if the cost is right (below $200,000 per unit). Each earth station will require one or more PBXs, and this is only a fraction of the market.

Another important item is the sophistication of the PBXs. The shift to computer control and the use of software have come about suddenly. As late as 1971, GTE was offering a model designed in 1937. With new technologies, the life cycles of PBXs quickly dropped from over 20 years to about 7.

The business opportunity, the cost, and the sophistication, in turn, influence office automation. The office of the future will have a central

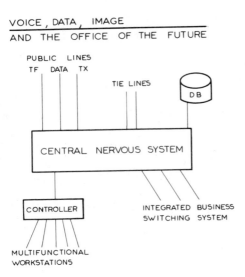

VOICE , DATA , IMAGE
AND THE OFFICE OF THE FUTURE

FIGURE 5-1   The central nervous system of the office of the future is the Private Branch Exchange (PBX).

nervous system (Figure 5-1) into which will feed a number of information transmission, switching, and storage services. Microprocessor-based switches can serve:

1. Line control functions

2. Message (voice, memo, letter, document) input (store and forward)

3. Message delivery

4. Dialogue communication (human–human, human–machine, machine–machine)

5. Telex and text implementation

6. Pure data handling (e.g., with an X.25 interface)

7. Facsimile exchange

8. Viewdata (interactive videotex)

and, in general, multifunction workstations.

The technology is available to assist an office worker with telephone, interphone, facsimile, teleprinter/telex, data terminals, viewdata terminals, word processors, and specialized terminals. But these services must be integrated; they cannot continue on separate networks. Separate networks, spoil wires, and setups increase the amount of needed maintenance and end up by reducing the comfort and efficiency of people at work. The needed integration can easily be assured through proper interfaces (Figure 5-2). One of the most valuable facilities this

WITH   PROPER   INTERFACES   WE   CAN   OPERATE   AN
INTEGRATED   NETWORK

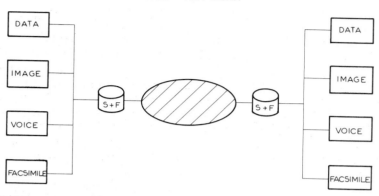

FIGURE 5-2   A store and forward capability (S + F) frees sender and receiver from the constraints of asynchronous communications, optimizes the use of the facilities, and prepares the ground for the integration of voice, text, data, and image.

interface can provide is store and forward because by dissociating sender from receiver, it permits optimization of resources.

Various possibilities exist for using such a system. Perhaps the most important is its autonomous usage, mastering the text and data flow of the office and channeling the information through public packet or circuit-switching networks. This is like one of the key objectives of the automated office: getting rid of paper, whether this paper is the result of a data, voice, or some other type of communication.

Dedicated computer-supported PBXs can be tuned to answer the need of merging message and voice facilities, thus offering the following advantages:

- Decoupling of sender and receiver, so that there is no need for both of them to be present
- Assuring electronic speeds and geographic independence
- Using computer technology to aid in composition, storage, retrieval, and reading

What is more, such solutions can operate within different environments: exclusive station-to-station calls, primary addressing with copies to other stations; broadcasts, with messages sent from one station to a distribution list; document distribution; and teleconferencing (group participation).

INTEGRATING DIFFERENT SERVICES

The knowledgeable reader will appreciate that no single device can answer such a potentially broad range of services as a PBX. We need tools characterized by flexibility and modularity. Flexibility can be assured through the software with which a computer-based PBX will be endowed. Modularity must be a built-in characteristic. Table 5-1 suggests five ranges of PBX capabilities I have found useful in my work.

TABLE 5-1   Ranges of computer-based PBX.[1]

Range of Supported Posts	Average Number of Ports	Projected Market
8–15	12	Sales offices, branch offices, professionals
30–40	35	Sales offices, branch offices, professionals
80–120	100	Satellites
200–300	250	Master
500–1000	800	Master

[1] The PBX can be miniaturized, incorporate modems, and in the lower range basically constitute intelligent telephones. These will be fundamental components for the office of the future.

In Chapter 6 we will follow an orderly approach for choosing a PBX responsive to user requirements. Still, the best machine that money can buy (as we have so often emphasized) would answer only some of the needs. Many more needs relate to preparation, and present-day experiences pinpoint several potential problems:

1. Machines fail, as we all know, and their users have not always known what to do in time-out.

2. Procedures on how to help a user in distress are still relatively undeveloped.

3. Manufacturers have not yet tooled up on how to train the end user on all modifications when they happen.

4. Users are still in the process of gaining experience on how to saturate the topology without increasing costs.

Clearly enough, equipment under development will be able to meet such needs, and we do know that the near future development of the computer and communications complex will incorporate the concept of the wired city, which will respond to a great many of these requirements. A different way of making this statement, although the office environment is complex enough in itself, is that a single paper-intense company presents a much simpler applications environment than does

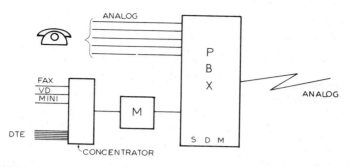

FIGURE 5-3   Use of a concentrator to minimize the number of modems neces-sary for the digital terminals: viewdata sets, minicomputers, and generally data terminating equipment (DTE).

a nationwide approach, and the benefits can be much more immediate and visible. We have the technology, but the procedural and end-user studies are not available to help in projecting, evaluating, implementing, and maintaining future systems.

Let us look at an example to document these conclusions. Figures 5-3 and 5-4 identify two solutions involving the use of a computer-based PBX on analog public lines. The alternatives being examined are basically use of a concentrator to minimize the number of modems necessary (Figure 5-4) against the simple attachment on the PBX of each device with its modem.

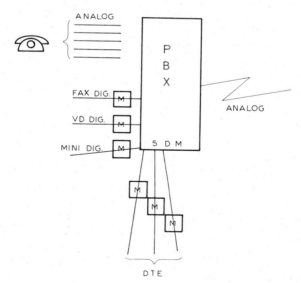

FIGURE 5-4   An alternative approach: direct attachment to the PBX, through modem, of all DTE.

Technology can support both options, but which one is better within a projected office automation environment? In a leading financial institution faced with this choice, the experts' advice was: Do not use a concentrator if the application is local; it is preferable to use more modems. The main problem is that switching takes place as a group; there is no individual independence of the devices.

The choice is not as easy as it sounds if we consider that the alternative also has its strong points. A concentrator solution permits more efficient use of facilities. It can make sense if digital stations are remote and if there is a mirror image of it at the distant end.

Further still, the solution is different if we are talking of digital voice transmission (PCM) technology. A dataset facility can then be provided between the PCM-based PBX and the analog lines (Figure 5-5). This evidently has a major impact on the choice of PBX equipment on equal or almost equal financial terms.

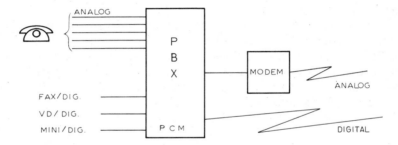

FIGURE 5-5   A digital technology, PBX requires no interfacing modems at the DTE side and towards the digital lines, but a D/A converter is needed towards the analog lines of the network.

Note: In the case described above, we did discuss with the experts the matter of the concentrator. Their reaction was that if a digital concentrator is used, there is no voice capability to be added, and for a post-concentrator we need modems. To the contrary, if a PBX/PCM technology is chosen, it will provide voice communication and, as stated, will not require modems at the digital station ends. We repeat that a PBX with SDM technology needs modems prior to PBX.

This is the current state of the art. There is, however, a Department of Defense–sponsored research project with AT&T for UNICOM, a unique, integrated interface able to use both digital and analog transmission capabilities. The goal is to develop a single PBX doing both analog and digital handling.

A COMPUTER-BASED PBX

To get a better understanding of what a PBX is and can do, let us return to fundamentals. Much effort is focused today, as we said, on developing sophisticated PBXs. Combining computers, memories, and switches makes PBXs able to:

1. Produce itemized, automated billing procedures, to allow the identification and management of toll calls.

2. Combine daytime voice-grade communications circuits into wideband data channels for nighttime high-speed information transfers.

3. Proceed with electronic mail (including office memos).

4. Combine voice channels into a wideband audio/visual conference circuit, including the swapping of visual information, such as slides and flipcharts.

Both the internal and the external calling capacity of the PBX system must be carefully considered because many business operations run a very high ratio of internal station-to-station dialing and a low-capacity system will not handle the required traffic load.

It is, therefore, important to examine the number of trunks and to establish attendant central office facilities used for outside connections, as well as the number of junctions (or links) which are used to set up internal calling paths. This study must be done within the perspectives of a well-established applications environment.

To understand the services computer-run PBXs can offer, it is necessary to introduce the subject of time-division switching. In a time-division switching network all connections are made via a single common bus called a "time-division bus." Every unit (line trunk) that requires a connection with another is provided with a port circuit. All port circuits have access to the time-division bus through a time-division switch. [When two ports require connection, their time-division switches operate simultaneously at a very high frequency (16,000 times per second). This technique, called "speech sampling," allows many simultaneous connections over the time-division bus. Each connection is assigned a time interval, the "time slot," and the number of time slots identifies the number of simultaneous connections among ports.]

The next critical issue is *circuit packs*. The system elements that we consider in Part Three—lines/trunks, switches, memory, and control—are contained on plug-in circuit packs. Each line circuit pack contains a number of lines, for instance, four. But the assignment of station numbers to line circuits is flexible.

The system memory is contained in circuit packs which provide the call processing functions. The circuit packs are held in small frames called "carriers." Within each carrier, the circuit packs are plugged into positions: the "slots." Every circuit can be addressed by, say, a five-digit number which tells its location by carrier-slot circuit.

There can be three types of carriers in a modern PBX system:

- Line carriers
- Trunk carriers
- Control carriers

The line carriers contain station lines. In AT&T's "Dimension," for example, a total of 52 to 64 lines are provided. The trunk carriers contain slots for 16 trunk circuit packs. The control carrier includes the processor, memory, control circuitry, data channel for attendant console control, and traffic measurement outputs.

This brief description of the technical issues connected to a PBX, and the ranges it brought in perspective, helps explain why the hardware configuration will depend largely on the services to be offered. Among possible services, we distinguish:

- CCSA
- CCIS
- Picturephones

Common control switching arrangements (CCSA) permit any unrestricted telephone station to call any other internal or external system station by using the standard seven-digit number. Alternate routing is a feature of CCSA service. The interfacility, alternate-routed calling paths are accomplished at the telephone company central office level, not on the user or subscriber's premises.

A system of interest to large-scale telephone users is *common channel interoffice signaling* (CCIS). Typically, this technique employs common channels to carry all interfacility signaling instructions: dial pulses, on-hook (idle), off-hook (busy), and so on, between two switching centers. CCIS replaces older methods of interoffice signaling such as "in-band" and "out-of-band" techniques. The former transmits signaling data within the normal conversational voice bandwidth. Its shortcoming is that false information may be transmitted due to unique tone or noise combinations set up in the talking path.

Out-of-band signaling techniques placed the interoffice data in special channels, generally adjacent to and immediately above the voice path. To preserve interchannel integrity, out-of-band signaling requires very efficient filtering or greater "guard band" separation between channels.

Picturephones are conventional CRT I/O terminals. With modern technology those employing television raster techniques using standard ASCII codes appear to be the best suited to business applications. They operate with a host processor and therefore perform a wide variety of applications, but picturephone switching still presents many problems at points through the carrier's networks:

- Most conventional central office switching systems simply will not handle the additional bandwidth employed in video transmission.

- The implementation of a video service will require a secondary, wideband switch to be made available to route the video portion of the call together with the voice. .

- Conventional wires associated with subscriber loops are not well suited to the transmission of images, whether in analog or digital form.

- Image transmission requires a bandwidth corresponding to 300 voice-grade lines.

Another key feature of modern, computer-based PBXs is memory. A lot of supports can be used, the best ones being magnetic disks, although the first computerized exchanges were equipped with tape cartridge. Typically, the memory device will provide off-line storage for information that is not used during normal system operation. For instance:

1. Load the program during initial installation or additions.
2. Store infrequently used programs (maintenance programs, traffic studies, and so on).
3. Reload the program after the memory contents have been destroyed (due to a power failure, for example).

Computer-based solutions need regular maintenance. A maintenance and administrative panel is the interface between telco personnel and the system. It usually contains a pushbutton data entry pad, an operational display, and indications for alarm and operational status. The various administrative and maintenance procedures required can be activated from this panel.

We have also spoken of software. Computer-based PBXs contain software for basic call processing, lists of available features, and translation information. These are referred to as "feature programs" and are largely dedicated. Associated with each feature program is a memory configuration that fixes the capacity of the program.

Furthermore, translation information (also contained in memory) includes line and trunk assignments, station class of service information, and other variable assignments within the system. This provides the capability for assignments of all features exactly as the customer wants them to operate.

KEY TELEPHONES

A special class of very small exchanges is that of *key telephones.* They answer the need to handle more than one line at a telephone location without the use of a switchboard, yet some form of simple switching is necessary.

The most widely used key telephone system today is the six-button telephone commonly referred to as "key equipment." To switch a call from one phone to another, the system must be equipped with hold relays, one for each line. These are activated by the red "hold" button, usually the first button on the left. The user answering the phone and wishing to hold that call, either to confer or to depress another line button, pushes the hold button, which activates the hold relay on that line.

Key telephones are also made as "one-button sets." They look like regular phones but have a single rotary button which allows the phone to be connected to either of two lines, often with a semiautomatic hold feature on the first line. "Call directors" are merely large key phones. They are furnished in 2- through 30-button sizes, arranged in columns or modules of 6 or 10 each. Hence the 18-button type has three modules.

The button associated with each exchange line is called a "pickup" button. Improper or unintentional depressing of pickup buttons is responsible for losing or disconnecting many calls. The caller who is "on hold" suddenly hears the dial tone and realizes that the connection has been broken.

To make key phones easier to use, telcos supply light signals under the buttons. One feature is a light that winks when a line is "on hold."

Key phones are used in conjunction with PBX systems when it is necessary to have several lines terminating on one phone in an office. Generally, key systems are best suited for small professional and retail sales operations. The advisability of their use depends on:

- The number of instruments used

- Lines available for personnel to intercept calls

- Optional features such as dial intercommunications between stations

To recapitulate: the key set is the familiar multibutton telephone station. It may be used to expand the internal flexibility of a telephone system, but care should be exercised to ensure that features associated with the PBX are simply not being duplicated unnecessarily.

This care is even more necessary as we move toward intelligent lines that can be managed centrally, through computers at the telephone exchange level, and which will integrate at the station (terminal) set a good many features, including interactive videotex, teletex, and store and forward capabilities. Such evolution may one day change the telecommunications architecture we are discussing in this chapter. Until that time intelligent telephones should have an excellent market in the private user field (households and professionals).

CENTREX

Centrex telephone systems have been originally designed to assure a significant degree of flexibility. They generally feature all capabilities provided by the PBX, permitting direct inward dialing (DID) of calls without intervention by the user's switchboard attendant and identifying outward dialing (IOD), which allows a long-distance toll charge to be assigned to each internal station.

With the development of computer-based PBXs which are polyvalent and offer services originally featured only through a central location, Centrex may be entering in disfavor and eventual decline. Many of the recent PBXs offer facilities which were initially available only through Centrex, such as station billing features, and more.

A computer-based PBX can also be used to emulate the IOD features of Centrex. So there is an option between requesting the telco to provide magnetic tape or hardcopy statements of the monthly long-distance itemized bill and keeping track of the expenses on the private branch exchange. Alternatively, telephone company records can be compared against the recording medium employed by the computer-based PBX to determine the validity of the carrier's bill.

Internal programs can also be generated to assign long-distance charges to individual stations, departments, groups within departments, or general overhead expenses. In different terms, PBX equipment or facilities centrally located on public utility premises may enhance cost control.

Centrex is a local telephone company service offering, not something that can be bought from a private equipment manufacturer. The private interconnect system can be designed to support Centrex, but these arrangements require detailed coordination with the local telco.

Some of the Centrex service, such as the direct inward-dialing features, can be offered by many of the larger, stored-program PBXs, but features associated with the assignment of trunks through DID must still be coordinated with the carrier. Incorporated with the PBX may be hardware used to emulate another feature of Centrex, the automatically identified outward dialing (AIOD).

PBX features which can supplement Centrex include automatic call distributors (ACD) equipment, which distributes incoming traffic among attendants, receptionists, or clerks, and is widely used in large-scale systems such as airline reservation offices or metropolitan police departments to ensure an evenly distributed workload.

Other features associated with ACD equipment are supervisory override functions, "forced release," automatically cutting off an incoming call if it exceeds a certain time period to prevent jamming of switchboards, and voice-or-music-on-hold. The last is employed to inform the caller that his or her call has entered the system and will be acted upon.

A good comparison is to say that Centrex has features similar to those of PBX, plus some additional provisions that give Centrex stations characteristics of both a business individual line and a PBX. Such facilities are offered at different levels of sophistication:

- *Centrex I* assures station hunting, station restriction, attendant-controlled transfer, direct inward and outward dialing, and individual station billing.

- *Centrex II* provides the foregoing facilities plus station dial transfer, add-on conferences, consultation hold, and trunk answer from any station.

- *Centrex III* guarantees all of the foregoing and call forwarding (from a busy or unanswered line), remote call pickup (the ability to answer on another extension), speed calling (abbreviated dialing), and touch-tone dialing.

AT&T provides Centrex services in two different ways, depending on the plan: "Centrex C.O." and "Centrex C.U." The former is an internal name used within telcos only to indicate that all equipment except the attendant's position and station equipment is located on telephone company property. The latter is also an internal name used to indicate that all equipment, including the dial switching equipment, is located on the customer's premises.

In a nutshell, Centrex has the following advantages:

1. *Fast inward service on most calls:* Calls go directly to the station user and an operator is involved only when the caller does not know the station number.

2. *Reduced payroll and related administrative expenses:* Fewer operations are needed and the internal accounting of calls is unnecessary.

3. *Modern attendant positions:* The attendant handles relatively fewer calls and uses the latest equipment.

4. *Complete communication flexibility:* The system has all the features of dial PBX service plus inward dialing and station identification on long-distance calls. Each caller is identified with his or her own telephone number.

With Centrex, a minimum of operator assistance is needed. Station users can transfer in-dialed calls, thereby relieving the operator of this duty. Centrex II is particularly attractive to those customers who have people working after the attendant position is closed. Centrex III offers added features, but there are also negative points to consider.

Centrex is costly, particularly from the standpoint of communications delays. It is subject to shared facilities, and can offer only limited access to the caller, who is more likely to receive busy signal delays, necessitating re-dialing of the number.

The difficulties previously experienced at the switchboard have not been eliminated, but rather increased, by placing them on each station user's desk. Thus Centrex is primarily of value when:

- The called party is always immediately available to accept the outside caller's call.

- Internal communication travel is nominal.

- Individuals receiving direct outside calls have a low volume of external communication requirements.

- Total communication requirements do not exceed the system's capacity to handle peak load volume.

Let's recapitulate. Centrex is a group of service offerings tariffed by the telephone company. There are two principal features that differentiate Centrex from PBX: direct inward dialing of incoming trunk calls to the telephone station without going through an operator (DID); and individual station billing for outgoing central office calls (IOD).

The equipment used by the telephone company to provide Centrex service is often the same as that for a computer-supported PBX. An important point to remember, however, is that with Centrex service most telephones must be "covered" at all times during the business day. All calls do not come through the operator, and it is bad business to have a correspondent calling an office and receiving no answer.

Chapter 6
CHOOSING A PBX

INTRODUCTION

We have defined the *electronic switching system* as the heart of office telecommunications and a determinant tool as to office efficiency, speed, and reliability. "Electronic" means that it is computer controlled; the computer helps to determine the number of facilities the PBX offers. Compared to their electromechanical counterparts, these systems are faster, more reliable, cheaper, and better. They also make feasible the effective handling of text, data, and image.

The new PBXs are a significant improvement on what we have had until now. This statement fully appreciates the fact that the most noticeable benefit any product can offer is that it will materially improve the user's productivity and ultimate profitability. The direct route to the top in any organization is through the management discipline most crucial to success—profits enhancement—that is what telematics should be designed to do.

Planning private telephone systems is a complex engineering task and relatively few are actually well planned. Because of the exclusive nature of the telephone business, until recently independent opinions were scarce.

In many organizations the telephone system grew unchecked, and the bills for basic service and equipment increased right along with the number of lines, stations, bells, buzzers, and miscellaneous features. In most organizations, responsibility for the telephone system rests, on rather vague grounds, with the building maintenance and administrative services. This diffused responsibility negates internal control and perpetuates the problem.

It is not unusual therefore to see that little, if any, in-depth understanding of the telephone system exists. Yet technology moves ahead at an unprecedented (for telephony) speed. The last 10 years have seen significant strides in private telephone exchanges: fast-speed dialing, touch tone, in-dialing with immediate access, means of dictation, follow-me (indicate through PBX where you can be reached when away from the office), short-code dialing, and conference facilities. Automatic ringback when the station is free, an add-on feasibility, an easy way of moving subscribers, and so on, are other facilities offered by the new PBXs. Because different telephone exchanges support various features, and common ones are given better solutions by some systems than others, the choice is complex. (We will demonstrate through a case study.)

But design-supported facilities are only part of the problem. Maintenance is itself a key issue, and diagnostic/maintenance capability (loopback and remote diagnostics) should be a major decision factor. The same is true of remote testing; the ability to read from a remote location the whole buffer of an exchange; and the ease in conditioning requirements (no forced cooling), power consumption, and so on.

The facilities, as a whole, must also offer the possibility for the end user to change his or her service requirements: for instance, to incorporate abbreviated codes, to support group-hunting arrangements, and to calculate the use of a station. Yet with the exception of cost accounting, the technological features often escape management's attention. Management knows only that service costs are so much per month and long-distance charges are always increasing. Beyond this, there may be various opinions regarding the quality and general limitations of the internal telephone system, but few solid procedures exist regarding the choice, management, and ongoing evaluation of the telephone plant and resulting operations. No criteria exist regarding the telematic services within the organization.

A DIMENSIONING STUDY

A PBX system is characterized by the method in which it handles interoffice and outgoing calls. With the manual board, an intercom connection is made by plugging two extension lines to a set of jacks in the

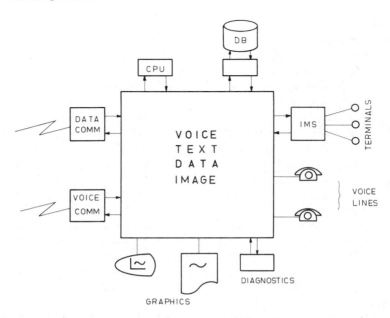

FIGURE 6-1   A PBX system supporting text, data, voice, and image and able to act as a gateway between the public network and the integrated voice functions.

board, connecting and ringing the called station manually with a key. When the conversation is over, light signals tell the operator to unplug the cord ends, which frees both lines for other calls. This is time consuming and requires full-time operators. In contrast, a computer-based PBX works without operator assistance for outgoing and intra-office calls. There are different levels of sophistication, the broadest possibility being shown in Figure 6-1.

Typical carrier-provided PBX systems were designed to meet the requirements of many average users and were tariffed accordingly. If a user did not fall into the definition of "average," or needed some special features, he or she was quite often forced into paying for more capabilities than required. We addressed this subject in Table 5-1, which presented some suggested PBX ranges.

So one of the advantages of a good private branch exchange is modularity. This is important because at the current state of the art, growth is difficult to predict. If we improve our communications system, the users tend to make greater use of it; the result is added intra-PBX and trunk traffic.

The right technical study starts with dimensioning and requires adequate planning. Confronted with the lack of quantification as to the exact nature, extent, and evolution of the user's telephone services, the manufacturer often advises: "Make the switching center bigger than you need today." That's wrong. The right way to go about a dimensioning study is to analyze the use of switching.

Switching is a reduction stage between terminal and destination. The reduction factor must be studied carefully; otherwise, we will create a bottleneck or pay for nothing. When talking of dimensioning, one must necessarily refer to saturation and its opposite: the capability of processing throughput traffic.

To dimension the system properly in terms of throughput capability, we should start from two different ends, remembering that the results to be derived must be matched.

At one end we look at time series, not only extrapolating on historical use trends, but also (and primarily) counting tendencies as more facilities are added: computers, word processors, intelligent copiers, viewdata, and teleconferencing, for example. Figure 6-2 presents this approach, reflecting the millions of telephone users as projected by four European PTTs.

At the other end we do what most companies consider the right thing: distribute a need questionnaire to all interested parties. Let me remind you that the obvious solution does not always work. Apart from the fact that many people dislike answering questions, the ultimate user normally finds it difficult to imagine what he or she will need in terms of services with a facility radically different from the current one. (At the 1979 NCC, reference was made to a 1975 National Bureau of Standards study which established the need for DP 500 terminals by 1986. The implementation was done in 1977. Two years later, June 1979, the actual installations exceeded 500.)

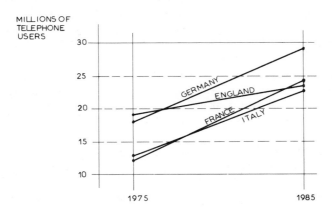

FIGURE 6-2 Current and projected growth in millions of telephone users in the four leading Western European countries.

Statistics are important to help choose one of the vital PBX aspects size. Initial size requirements must be defined in terms of:

1. Number of station lines
2. Number of trunk groups and types
3. Total trunk quality
4. Number of attendants' consoles required
5. Traffic-carrying capacity/voice
6. Traffic-carrying capacity/text and data
7. Projected growth
8. Reliability

Modularity can be the answer, as traffic volume is the limit to providing greater facilities (attaching more users and add-ons to the same PBX). But modularity also can have its limits: quality service to the connected stations is based on holding time for station-to-station calls at 40 seconds and for outgoing calls at a maximum of 90 seconds. We know how to calculate in erlangs.

The PBX is a switch, and the basic philosophy in designing a switch is to provide for reduction states between

- Input
- Lines, and
- Output

Other problems, as well, require attention in the planning stages. They relate to the work environment and its evolution. For instance, with financial institutions the problem is more complex because the general management, some detached central offices, the branch offices, and the computer center are usually spread over the city, yet management may wish to link them through the same PBX at the headquarters level.

The question must then be asked: Where are the operations located? What's the topology? the distance? the link to the public network? Still, the structural issues are only part of the picture. To plan a PBX service, we must first understand what this service is and what it can offer.

Reliability brings into perspective the need for communicating with the system. Computer-based PBXs offer a number of possibilities, including fault reports, to inform a maintenance staff. It is a good idea to ask the manufacturer for statistics. In a recent selection case a manufacturer documented an assurance of 1.4 faults per year for every 100 subscribers, and the machine provides rerouting.

Let's have another look at reliability: "time-out." Can total system outage be tolerated? If the answer is yes, money can be saved by buying a system without redundancy in the common control. If the answer is no, redundant common control is mandatory.

Design can assure that system-wide time-out is practically nonexistent, given built-in duplication of process cards, power supply, and so on. If a system outage can be tolerated, money can be saved by avoiding battery support, if not, batteries are needed and their reserve capacity must carry the PBX load for the longest expected power failure interval.

INCREASING THE CAPABILITIES

As dial capabilities have been gradually implemented, the PBX enabled internal stations to dial each other without attendant intervention and to gain controlled access to trunk lines. External trunk lines can be provided to meet a wide variety of specific applications, including toll restrictions, toll diversion, two-way trunks, one-way incoming or outgoing trunks, and so on. The same can be said of internal services.

Operating facilities may include, for instance, the chief secretary system, which significantly helps management. Say that we have two principal offices, and the chief executive officer is traveling. Should the facility extend to all offices and possible sites of visit? If yes, we need efficient protocols for facility handling. The chief secretary capability, typically, would include the facilities for diverting calls and providing a system profile for in-dialing (public network to station). Administrative prerequisites would add the need for counting.

Externally, the size of the private network is a major problem. Size and facilities bring many extras to the public station-to-station system. Interlinking dramatizes the fact that standards vary not only from country to country but also from manufacturer to manufacturer within the same country. (I was recently in a Phoenix resort hotel and called a local computer manufacturer early in the morning. As soon as the operator at the factory tried to transfer the call to the requested station, the line died. This happened five times! At the last trial, the factory switchboard operator asked: "Are you calling from 'The Pointe'? That happens often. The two telephone exchanges cannot work together." And all that nice talk about compatibility?)

Hence, a vital consideration in selecting a PBX is how it will interface with the world. Is this PBX isolated, or is it part of a network of several PBXs? Are there special voice or data communications circuits that must interface with the PBX? Would it be surprising to find that existing signaling or supervision arrangements are not compatible with the PBX?

TABLE 6-1   Private branch exchanges.

Features	Simple PBX	Complex PBX	Centrex
Attendant console	X	X	X
Direct outward dialing	X	X	X
Direct indialing	—	X	X
Station-to-station calling	X	X	X
Station hunting	X	X	X
Station billing	—	X	X
Call transfer/attendant	X	X	X
Vacant number intercept	—	—	X
Reception from outgoing calls	X	X	X
Power-failure transfer	X	X	X
Assigned night answering	X	X	X
Attendant of camp-on	—	X	X
Indication of camp-on	—	X	X
Attendant conference/consultation	—	X	X
Call pickup	—	—	X
Call transfer individual	—	X	X
Call forwarding—no answer	—	—	X
Call forwarding—busy	—	—	X
Add-on conference	—	X	X
Trunks answer from many stations	—	X	X
Abbreviated dialing	—	—	X

The projected layout must be studied as carefully as the external technical features to assure interconnection. The same is true of the facilities to be incorporated into the PBX system: for instance, attendant trunks and information trunks. Table 6-1 identifies the facilities offered by simple and complex PBX systems and a Centrex service such as the one AT&T provides, which is based on computer equipment installed at the central office.

Over the next decade we may see more features added. Videophones on an *internal* station-to-station arrangement may be desirable. Higher switching speeds will be featured, but here certain human factors come into play. In normal human-to-human calling modes, higher switch speeds surpass human ability to dial or answer a telephone call. In terms of the total call cycle:

- Dial
- Switch
- Ring
- Respond
- Answer

the human being is the greatest limiting factor.

Furthermore, calls external to the private interconnect PBX will still be limited by the ability of the carrier central office to accept high-speed dial impulses, or alternate voice/data communications.

Attendant features is a critical issue. A wide variety of attendant console features are offered by various manufacturers. Needs must be identified, as they may influence the choice of console type and features. What would the attendant need? There are key-per-trunk consoles where each inter-PBX and central office trunk, as well as "Dial 0" and intercept trunk, has its own dedicated key appearance on the console.

Selection of the PBX must also involve examining the problems peculiar to electronic systems:

1. Vulnerability to lightning entering via trunk or station pairs and power lines

2. Sensitivity to static electricity

3. Grounding requirements

Other considerations include the manufacturer's recommendations for spares to support a given reliability level, the cost of these spares, the supplier's maintenance network, the technical backup the user will receive, and the systems assistance available.

In terms of cost, users should ask themselves how many PBXs of this type they plan to purchase. Although important, PBX cost is only one of the telecommunications costs organizations face, and as Figure 6-3 shows it is not the most significant one. According to European statistics, PBX acquisition uses 9% of the yearly budget, maintenance uses 6%, about 6.5% is paid to telco, and nearly 20% of the budget goes to the operators.

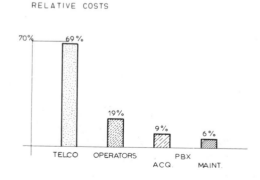

RELATIVE COSTS

FIGURE 6-3 Relative cost in telephony, including the implementation of private branch exchanges (data based on Western European tariffs).

AVERAGE COSTS BASED ON FRENCH, GERMAN, AND ENGLISH DATA.

The cost of purchase (or rental) of the equipment proper should not be the unchallenged factor of the choice. Proper planning of telephone services requires a careful engineering study and a factual examination of the system's features, including performance, reliability, and compatibility guarantees.

The user conducting the acceptance tests must develop an internal skill regarding the installation method, be familiar with service manuals and parts lists, obtain assurance on future field modifications (voice, text, data, image, and teleconferencing services), do traffic and usage studies, plan for disaster recovery, and educate the staff regarding telephony and its use.

There is no better assurance than a strong contract with the PBX manufacturer, and the same is true with computers. The contract must specify the reliability and availability guarantees and cover all issues relative to damage or penalties, including breach prior to delivery, breach after delivery, liquidation damages, maximum damage, and breach with consent.

THE CHOICE OF EQUIPMENT: A CASE STUDY

In early 1980 a leading European financial institution decided to purchase two PBXs to be installed at their headquarters office and a new computing center. Nine reputable telephone exchange manufacturers were called in the first round: GTE, ITT, Northern Telecom, IBM, Siemens, Alcatel-DIAL, Erickson, Telenorm (TN), and Philips. They were reduced to six in the second round, and one was chosen in the third round.

The local telco did not participate in the competition, but its advice was sought for four basic reasons: its experience in telephony, the interfacing of the PBX with the telephone lines, the structural connection between headquarters and the new computing center, and in the last analysis it is the telco which (according to local rules) will maintain the PBX equipment.

For our part, we outlined the prerequisites for a rational selection, given the bank's present and projected requirements in terms of voice, text, data, and image. The telephone company, as always in the last 10 years, had shown plenty of goodwill and collaboration.

Having spelled out the objectives and established the alternatives, the first round dropped three of the competitors mainly for technical reasons. (For instance, one of the equipment systems handles voice only.) Then a careful evaluation was made of the remaining possibilities. The comprehensive results are presented in Table 6-2.

TABLE 6.2 Results of the evaluation of PBX manufacturers.

	Integration Voice–Data	Technology	Modularity	Characteristics	Local References	Qualitative Remarks
PBX I	Not at the same module, but can be done through different modules	PCM, TDM	Modular	Multiuser remote diagnostics	8	Modular design permits specialization of modules (voice, data) and also backup possibilities
PBX II	No	PAM, TDM	Not very modular, three blocks	Limited remote diagnostics	12	Starts getting obsolete
PBX III	No	SDM	No, only at board level	No remote diagnostics	30	Starts getting obsolete
PBX IV	Yes	SDM	No, only at board level	No remote diagnostics; can be attached only to PBX of the same manufacturer	±15	Starts getting obsolete
PBX V	Yes	TDM	Modular	Remote diagnostics	1	Lack of experience in field maintenance
PBX VI	Yes, but does not support complex data load	SDM	Modular	No multiusers	3	Failures due to carpeting, temperature, overload, memory spillover

The following criteria were used:

1. Integration of voice, text, data, image
2. Technology (TDM, SDM)
3. Modularity
4. Equipment characteristics
5. Local references
6. Qualitative remarks

Technically speaking, the best equipment of the lot was PBX I: modern, efficient, modular, PCM–TDM based (hence advanced technology), able to work as Centrex (the larger model), multiuser, able to perform remote diagnostics (very important), able to specialize its modules on voice and data, and made by a major telephone company that knows its job.

PBX V had similar features, but the local telco did not believe the company representing it in this market was ready to give good support. (There were also some long-outstanding problems between the telco and that company.) Furthermore, the telco disregarded alternative IV on the ground that it is obsolete (which it is) and it advised against PBX VI, as half-baked equipment with plenty of technical problems.

Since the subjects of cost and data load are related, it was necessary to reexamine the estimated traffic load prior to making the next step in the selection. We agreed that two alternatives should be established: one a minimum configuration and one maximum. The minimum represented a fair increase on present conditions; the maximum was what the bank may need in seven years.

Solution 2 seemed best, and the work load was examined carefully. At that time, the bank had about 350 telephone stations but fewer numbers because two stations occasionally shared a number. There were also only a few DP terminals online, but more had been ordered and others were planned. The same was true of minicomputers. (By the next year the company expected to have at least six minicomputers linked to the central machine, and a lower two-digit number of terminals, all to be served by the PBX. Ten word processors were also projected to be hooked up.)

TABLE 6-3 Maximum data and voice load.

Location	Voice	Data	Voice and Data
Headquarters	250–350	50	400
Computer center	300–400	100	500
Total	About 600–750	150	900

To consider the maximum configuration, given the semiconductor revolution, the group decided to explore the possibility of having one terminal on every desk by 1985–1986. To a large degree this should integrate voice and text data. Table 6-3 summarizes employment statistics.

REPORT TO MANAGEMENT

The report to management stated: "It is evident that the initial installation should only answer requirements for the next two years. Beyond that time, the modularity of the PBX should assure that extra capacity requirements are answered as they develop. Let us also take this opportunity to stress the need that both the computer center and the extension of the head offices should benefit from the developing communications technology at the drafting board. After the construction is done, changes will be costly and inefficient."

The projected PBX features included an alphanumeric display for attendant position, attendant conference, attendant control of trunk group access, automatic callback, attendant lockout, automatic route selection, busy verification of station lines, and call forwarding for all calls.

There was a range of call facilities to be supported: hold, pickup, waiting services—attendant, originating, terminating—and a host of controlled restrictions:

- Outward
- Inward
- Station to station
- Terminal
- Manual terminating line service
- Termination
- Total restriction

Among the PBX features included in the final choice were direct inward dialing (DID), trunk group selection, executive intercom, executive override, loudspeaker paging, multiple listed directory numbers, night station service, and outgoing trunk queuing. Privacy and lockout were supported, as were recorded announcement interfaces, telephone dictation access, serial call, speed calling, station message detail recording, three-way conference transfer, and timed recall on outgoing calls.

Other projected characteristics included remote access to PBX services, trunk answer from any station, trunk group warning indicators, trunk verification by customer, trunk-to-trunk connections, and two-party hold on console. This should support our statement that for proper telephone services planning one requires a careful engineering study to provide a complete, detailed design of every voice station, text, or data terminal to be installed. The study should include the phone's location, the lines to be answered on a key set, and any other special features.

In negotiating the price of additions or deletions from the contract, we risked being in a weak bargaining position without such study. A factual examination of the features found some of the systems wanting, and without an analysis of system performance, we were not in a position to know if the PBX could handle the anticipated traffic rate. Accordingly, both the traffic-handling requirements and the rejection rates were included in the specifications.

Evaluating system reliability, we insisted that the supplier guarantee rates and criteria for expected service and downtime. This applied to individual lines, trunks, and the total system. In asking for a compatibility guarantee we sought manufacturer assurance that due to the rapid changes being effected by electronics, the new designs would not make system expansion incompatible with the telco network for at least the term of the payback period.

In conducting acceptance tests we applied a principle we have been using for years with computers: client-written tests should be run on the PBX before it is accepted, including the artificial creation of anticipated peak traffic conditions. In this particular case, tests covered possible disaster recovery, to answer the question: In the case of a complete disaster, such as fire or flood, what provisions are made to assure rapid recovery of service?

Competition was tough, and this kept the manufacturers on the defensive and they came up with reasonably good offers. Since cost was one of the primary factors but not the only one, we asked some quantitative questions about the cost of the equipment, the software costs, the service and installation cost (if extra), the hidden costs behind some manufacturers' moves and revised offers. The highest price offered for the two PBXs was around $600,000; the lowest was $320,000. Since the solutions were technically comparable, we were happy to settle for the lowest cost.

Chapter 7
THE COMING TELEPHONE TECHNOLOGY

INTRODUCTION

Microfiles, text/data warehousing, and databasing requirements at large are going to be the foremost subjects of the 1980s. Along with them, text, data, and voice communications are destined to play a vital role, yet the function of verbal telephone messages cannot be ignored.

The computer-based intelligent private branch exchange constitutes a vital link in the DB/DC network, its functions growing steadily. Tied to a PBX workstation used to integrate text and data handling with the voice telephone network for least-cost routing, PBX will be widely used as the gate to the communications network.

The internal IBM version of such a scenario can already demonstrate:

- The disappearing need to exchange telephone numbers
- The fact that some 15,000 people regularly use the network, averaging more than two transmissions daily
- The usage range, which includes company mail, memos, files, documentation, and programs
- The potential population (all 310,000 IBM employees) having access to the network

- The initiative to use public (not private) tie telephone lines
- One very important user perspective: beyond the terminal everything else is transparent to the user

The sort of market to be supported through the integration at the user's end of the text and data facilities can be exemplified through statistics. If all current types of electronic message services, including telex/TWX and Mailgram, are considered, the current spending by users on electronic mail is about $1 billion in the United States alone. This will be increased multifold by the end of this decade as business places at the end of the telephone line integrated terminal equipment.

REDEFINING THE PBX FUNCTIONS

Let us make the hypothesis of an online system with storage and forward capabilities which permits the user to send a memo by entering on a keyboard text and data stored in the system. The user can also leave recorded spoken and written messages to be transmitted to the other party's terminal immediately or when specified by the sender station. Through PBX support, the end user has only to press the appropriate function key. The other party, upon his or her return, sees the message and can immediately initiate the appropriate action.

Along this line of reasoning, Figure 7-1 divides the capabilities of a computer-based PBX into three major areas:

1. Switching oriented toward the external network (the environment the PBX interfaces)

2. Switching for the voice traffic relative to the industrial environment

3. Text- and data processing-oriented functions, including the interfacing of terminals, workstations, and the administrative duties to be performed at the PBX level

These three classes have similarities and differences. Switching, for example, whether oriented toward the external or the internal environment:

- Is characterized by process control-type applications
- Calls for very high equipment reliability
- Requires the ability to match line characteristics
- Implies a fast response (10^2 ms level) toward both the line *and* the user

The provision of storage capacity for these functions makes bubble memory use advisable for the switching functions, but from a design

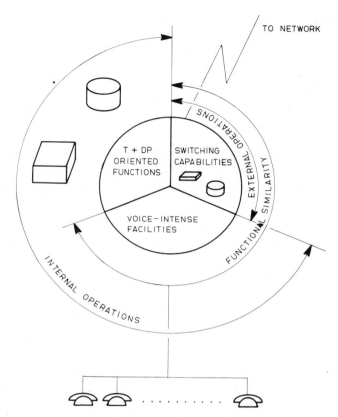

FIGURE 7-1   Division of the PBX capabilities into three major functional areas.

standpoint, these functions are much more linear and well known than those of the text and data processing (T + DP) part and the administrative chores to be supported.

To design the technical characteristics of the T + DP part, we must properly identify the intended application:

1. Group the requests to be posed at the switching part (housekeeping, whether internal or external).

2. Assure a dynamic allocation of resources.

(Regarding applications, it will be necessary, for instance, to instruct the PBX processor about a necessary change of parameters which may involve output lines, internal lines, and a number of allocation-related problems.)

3. Modify the end user's T + DP and telephony-related characteristics (long-haul access, direct dialing, and so on).

4. Make feasible system preparation functions (generate, customize, and so on).

5. Assure system maintenance (for the foregoing parts).

6. Provide an infrastructure for accounting and billing.

So far, emphasis has been placed on the enhancement of rather classical PBX functions, but a computer-supported PBX must also provide and promote new capabilities. Examples are:

7. Extend the accounting functions into statistics, with supporting issues such as traffic analysis.

8. Provide journaling facilities for T + DP and voice traffic and, by extension, voice mail.

9. Cable management capabilities for installation support.

10. Open to the end-user capabilities, including such fast-developing requirements as:

    a. Online data collection

    b. Control monitoring

    c. Site operation (open and close doors, and the like)

We can present the 10 facilities that have been enumerated (and any others we might wish to add to the list) as *functional subsets* in a matrix form (Figure 7-2) against a variety of *usage classes*. This matrix can subsequently become the relational-type organization of the PBX database to be assigned on disk (to serve all three main sections presented in Figure 7-1) and open to further expansion as the need demands.

Notice that to keep the system logic simple, we can nicely develop matrix families (up to 100 in the case of Figure 7-2) which are firmware-supported and still make finer divisions feasible within each family through a class concept: the members of each family of functions, while sharing common characteristics, can be further distinguished through proper identifiers.

This approach to a multifunctional PBX design, by abandoning old principles in telephone exchanges, breaks with the past. There was only one monolithic unit at the PBX and user levels, but new designs exchange the mainly physical known entities for new, presently less well defined functions which are primarily logical. Through this transition it has been possible to bring in the notions of:

*   Voice store and forward
*   Text and data store and forward
*   Image handling

FUNCTIONAL SUBSETS \ USAGE CLASSES	END USER TELEPHONY	RESTRICTED LISTS	DOCUMENT DISTRIBUTION	DISTRIBUTION LIST UPKEEP	CONTROLLED DATA COLLECTION	VOICE MAIL	DATABASING CAPABILITY	ACCESS CONTROL, DOOR OPENING	REPARTITION OF SERVICES
1. REQUESTS / SWITCHING PART									
2. DYNAMIC ALLOCATION									
3. MODIFY/END USER SIDE									
4. SYSTEM PREPARATION									
5. MAINTAIN-ABILITY									
6. ACCOUNTING / BILLING									
7. STATISTICS/ TRAFFIC ANALYSIS									
8. JOURNALING									
9. CABLING MANAGEMENT									
10. OPEN END/ USER EXITS									

FIGURE 7-2 Organization in a matrix form of possible PBX facilities which will eventually be microprocessor-supported.

- Storage security/protection
- Housekeeping and maintenance

To enhance functionality and aid in the need for definitions, we must structure the new requests into a formal discipline supported through *protocols* and *paging*. We must also provide the necessary detail to support security/protection capabilities as the services provided by PBX systems become increasingly widespread.

An example of the detail and formal definition required is the telephone number itself. In the past (and up to the present) the telephone number confused:

- Station and
- User

This practice is no longer acceptable. The telephone number must distinguish the right way:

- Station

- User, and
- Author

This calls for introducing capabilities addressed to various levels. The user and author must identify himself and herself (through badge, voice signature) at any station he or she uses to gain access to the PBX, and the latter must dispose the proper controls to check the request and assure authentication online.

Databasing capabilities are evidently called for to support this facility, but the original prerequisite is one of protocol definition. Formalisms are necessary to sort out another mix-up, that between

- Signaling and
- Communication proper

Here, ideas that we have applied with data communications, such as the header, can help, while other issues can further support the drive toward more sophisticated approaches in voice traffic. An example is the use of the trailer for error correction capabilities. Substituting the star distribution network through a loop concept (Figure 7-3) is a major system change to be enacted in implementing a PBX within an internal operating environment. This will do away with the need of the distribution board and exchange the physical port by the logical socket.

Supposing that the PBX is supporting 10 different processes and

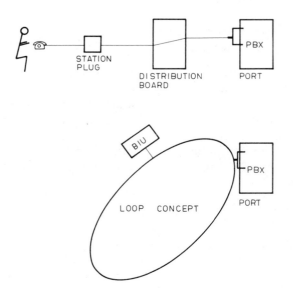

FIGURE 7-3   A loop concept for a local network architecture.

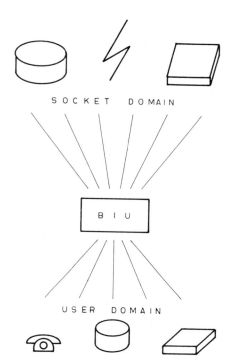

FIGURE 7-4 Possible service domain
of a bus interface unit.

has access to, say, five physical ports, the system will be running an
aggregate of 50 sockets. In a loop structure the *user domain* will be
connected to the interface units (BIUs), each supporting a defined
number of stations (Figure 7-4).

The BIUs will substitute *plug identification* in present-day systems,
it being the best means to formal definition of the end station. (A re-
movable telephone set can be hooked to more than one plug. In this
sense, it constitutes a better reference point because it is fixed.)

If the fixed plug is seen as the point that can be the most easily
identified in taking a local view of the internal environment, this
changes substantially when we consider a system view, which necessarily involves radio links and gauging devices (Figure 7-5). In a system
view, the logical identification is pushed further down the line through
the insertion of the concept of the mobile station.

The system view of the coming private branch exchanges will,
furthermore, include a proper emphasis on software: both the operating system and utilities. (A recent estimate calls for 50,000 instructions.) It will support user exits, handle parametric programs, and
make feasible cabling management through firmware and table lookup.
Telephony was very stagnant for 100 years; it has finally exploded.

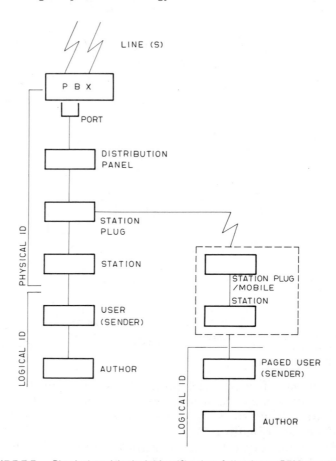

FIGURE 7-5 Physical and logical identification following a PBX connection.

NEW PROFILE FOR TELEPHONE COMMUNICATIONS

Not only are the mechanics of telephony subject to fast evolution but the dynamics are also changing. An impressive list of communications companies suddenly compete for the high-capacity voice and data communications network the modern enterprise needs to conduct its operations in an able manner. The top two dozen in the American market include:

1. American Telephone & Telegraph
2. American Bell (AT&T's new unregulated facility)
3. American Satellite (AmSat)

4. Satellite Business Systems (SBS)
5. Comsat
6. Cable & Wireless
7. MCI Telecommunications and its Execunet
8. GTE-Telenet
9. Tymnet
10. Graphnet
11. ITT/DTS (Domestic Transmission System)
12. ITT World Communications
13. RCA Americom
14. RCA Global Communications
15. Southern Pacific Communications (sold to GTE in late 1982 for $750 million)
16. TRT Telecommunications
17. Western Union Telegraph
18. Western Union International (sold by Xerox to MCI after the abandonment of XTEN)
19. Isacom (a resale carrier)
20. Contemporary Communications
21. Digital Termination Service
22. National Microwave Interconnect
23. Data Communications Corp. (DCC)
24. Local Digital Distribution (LDD)

Given their diverse origin and financial staying power, these companies do not appeal to exactly the same markets. Nevertheless, their market segments tend to overlap, bringing in competition. MCI and SPC favor the voice user rather than the datacomm user. (SPC took over the Data-dial service when Datran folded in 1976 but no longer offers that option.) Both companies are pushing their competitive offerings to AT&T's Message Toll Service, "and they are enjoying a significant market acceptance. This competition centers on the most profitable markets served by AT&T Long Lines."

The data communications requirements find their answer in the value-added, packet-oriented networks and in the satellite carriers. Between these lines, there are a variety of communications approaches that can be used, but comparisons can be difficult, since they will include such diverse offerings as Telenet, Tymnet, American Bell/Net 1, ITT/DTS, IBM's Information Network, American Satellites, RCA

Americom, Western Union, and SBS. (As of March 1982, 17 users of SBS services employed 52 earth stations, projecting among themselves the need for 174 earth stations by 1984.)

The prospective user of the new breed of communications services must consider a number of key factors:

- Service capability
- Technological features
- Geographic spread
- Cost

The services modern technology can offer should be analyzed into transmission efficacy, error-free operations, data integrity, and interface options. Every one of these factors presents conditions that can be expressed into discrete advantages and disadvantages. In turn, these characteristics will reflect on each commercial network, and they must be given due weight before a decision is reached.

Able, fact-oriented commitments necessitate a good grasp of the present and future data flow within the user organization. The preliminary research should develop such statistics as:

- The number and locations of the stations in the network
- Their technology
- The expected data traffic volume
- The evolutions of this volume
- Time-dependent delivery requirements
- Tariffs

System topology is a prime consideration, as value-added networks are usually constructed around a number of geographically distinct nodes that control data communications between user terminals and the host and also run the network.

Topology also plays a key role in local connections. Chances are that the terminals and CPUs at the user side will be linked to the value-added network (VAN; or value-added carrier, VAC) node through leased or public lines not supplied by the VAN. Such connections (particularly the dial-up lines) might reduce the VAN quality offering in terms of data transmission speeds, data integrity, and so on.

Furthermore, the VAN offering itself needs to be scrutinized in terms of the capabilities which it supports in adjusting protocols (S/S, BSC, packet), changing data from one character set to another, speeding data along the most efficient route, and multiplexing data from a variety of sources onto fewer telephone lines, but also in regard to services such as electronic mail and message switching.

In terms of tariffs, an important point to remember is that the expense of using value-added networks is determined by traffic volume and is independent of distance. To the contrary, the expense of using the old-line common carriers is determined by distance and, once the line is leased, it is independent of the degree of line usage.

TRANSFORMING THE CLASSICAL TELCO

A structural reorganization is now under way in the slow-changing telephone business. Telcos rightly project that since transmitting data and text may soon become as important as sending voice, the market for communications services will grow so fast that it may even surpass the size of today's voice business. There seems to be little doubt that American Telephone & Telegraph and General Telephone & Electronic have set their strategy to match this business opportunity.

For the smaller independent telephone companies, however, the "best way" for survival is far less certain. Not only is the VAN challenge difficult to meet, but also such new services as cable television and satellite-based communications threaten to take over their rural telephone franchises.

Take the most aggressive among them as an example of the struggle for survival. With $3 billion in assets, Continental Telephone is tiny in comparison with AT&T. Its revenues of $1.1 billion amount to just over 2% of AT&T's, but Continental has an aggressive management and a strategy for transforming itself into a broad-based telecommunications company.

Through acquisitions and joint venture investments between 1978 and 1980, Continental assured itself of these major pieces needed to become a supplier of advanced services to the business community:

- American Satellite Corp. (AmSat), a joint venture with Fairchild Industries Inc.

- Executone, a PBX marketing company which provides the base for a business in intelligent terminals and digital telephone switches

- Omni Communications, which gives Continental a hold in cable television

Some of these ventures will bring Continental into a head-on clash with the giants. While for the moment, AmSat and SBS are not directly comparable (SBS can relay information from its satellite to more than one location at a time, while AmSat is a point-to-point system), in the long run this will not be true. If SBS supplies its customers with a transmission capability that can be increased or reduced depending on what the customer needs, AmSat will have to offer a similar service to survive the heat of competition.

Indeed, ASC (American Satellite Corp.), jointly owned with Fairchild Industries, introduced several services targeted for data users:

- ADX (Asymmetric Data Exchange) is designed for bulk data transmission from a host CPU to remote terminals at 9.6, 19.2, 32, or 56 kbps, with a low-speed land-based return channel at 4.8 pbps.

- SDX Metroline is a multiplexed version of the SDX (Satellite Data Exchange) service and offers individual 56-pbps channels to different users through a common rooftop earth station located at one user's premises.

- DDX (Distributed Data Exchange) permits establishment of individual 9.6-kbps links to handle various types of traffic.

Voice, data, and facsimile are carried at rates to 9.6 kbps and eventually with the new 56-kbps wideband data service, at 56 kbps.

The Continental's management seems determined to capitalize on the advantages of being small. It can move faster than its competitors because big companies have greater overhead as well as resources. Smaller firms can turn business around faster. For instance, Continental is planning experiments to provide an electronic newspaper, an information retrieval service, and computer-supported phone books to residential customers via its phone lines.

Continental is also:

- Negotiating with several banks to market bank-by-phone services
- Talking to utility companies about offering a telephone-based energy management program

Its management is betting on the Omni CCTV capabilities, as sophisticated two-way cable systems with electronic mail capabilities might eventually enable the user to bypass the telephone network.

Yet it is just as true that economies of scale are hard for the smaller telcos to achieve, particularly if their telephone properties are dispersed. Much of the success of current plans will depend on the new markets. A stronghold in data communications will be a function of the transition from a regulated monopoly to an aggressive, marketing-oriented firm.

One of the most important ingredients in such a transition is the training and attracting of able management. Another key factor is the ability to position the available resources to meet the forces of the 1980s. To list some of Continental's weak and strong points:

- Its telephone sets are used an average of less than 15 minutes per day.

- Only 1% of its customers used advanced features such as "call forward" and "call holding."

- Only 6% of its subscribers have pushbutton phones, versus 40% at AT&T.

- Its growth rate is about 50% higher than AT&T's because rural areas are becoming urban.

- Continental has 50% more of the needed digital switching systems installed per capita than AT&T.

No firm, not even the giants, can bet in all directions and win. Continental and all other challenging telcos must decide precisely which market they are after and center both management skill and financial resources accordingly.

STRATEGY AND STRUCTURE

GTE is another telco that reached into other sectors of the electrical engineering industry in the postwar years. The "number 2" regulated telephone utility in the United States, albeit with a much smaller share of the market than AT&T's, the 1979 purchase of Telenet brought GTE in the foreground of the value-added carriers.

GTE Communications Network Systems was organized for global marketing of data and voice communications services, and the results have been good. In 1980, GTE Telenet introduced "Telemail," a computer-based mail service which enables users to

- Send,

- Receive, and

- File messages

anywhere in the United States, using a variety of data terminals and communicating word processors.

GTE plans to expand Telemail into a high-speed multimedia information distribution system with databasing capabilities supported at the switching centers. To overcome the incompatibility of different vendors' equipment installed at the user sites, Telemail assures speed, code, and format conversions.

GTE Telenet plans to develop other network interfaces for batch terminals, fax, intelligent copiers, and micrographic units, eventually accepting and delivering voice messages through store and forward techniques.

Management also plans to deploy a satellite-based packet network able to transmit data at megabit speeds, involving satellites and GTE's own earth stations on the same sites as its major central offices. Among the important transformations is the project's increase in the transmission rate of portions of its existing ground network from 56 kbps to 1.5 kbps, helping reduce end-to-end network delay from 200 ms to about 50 ms.

GTE's ultimate aim with its Telenet services, its Interactive Videotex connection, and its license to produce personal computer-type terminals is to bring database services to the home. Its current priority, however, is the office. When that curve reaches a point where it reduces terminal costs, the residential market will begin to develop.

The financial staying power Telenet now masters (through GTE) moves the company a step forward in its long competition with Tymnet. Both are packet-switching carriers: the former an X.25 disciple; the latter follows its own protocol, which maintains its more efficient accounting for the message stream of its clientele.

Tymnet is the first value-added network established by Tymshare, a service bureau organization operating coast to coast. The Tymnet system accommodates up to 32 high-speed synchronous input/output channels carrying data at rates up to 56 kbps (or up to 256 slower asynchronous channels).

In the United States, Tymnet has introduced a new low-cost electronic mail service, "TymeGram," intended to handle high-volume first-class mail.

The International Switched Interface System (ISIS) permits the interconnection of a variety of terminals and handles different protocols: asynchronous and synchronous interfaces, X.25, and so on. This makes it feasible for network users to access Tymnet over dial-up or private lines.

User terminals are connected to terminal interface processors (Tymsats) which balance transmission speeds and carry out code conversions automatically.

One of Tymsat's major competitive advantages is that it can be accessed from over 30 countries (although local restrictions make some of these connections difficult or costly to implement). Interconnection is provided under agreements between international record carriers and the PTT of each country. Canada's Datapac network supports direct access to Tymnet.

RCA Americom's private-leased channel rates are determined by:

- Equivalent voice-grade line charges
- Station termination
- Line terminal charges

Several agencies are using RCA Americom links, including dedicated earth stations, for data rates to 1.5 mbps. A 50-mbps link is projected for use in connection with the Space Shuttle program.

Western Union Telegraph has transformed itself into a satellite carrier. It also offers a variety of communications facilities for users requiring point-to-point and multipoint transmission of voice and data:

- The "Low Speed Channel Service" is designed to carry data at rates of 75, 150, and 300 bps and is offered in over 360 U.S. cities.

- The "Medium Speed Channel Service" is designed to carry voice as well as data at rates of 1.2, 2.4, 4.8, and 9.6 kbps.

A DataCom service permits users to bundle their intercity transmission channels using time-division multiplexers, with transmission speeds from 75 to 1200 bps.

InfoCom users can implement their own private networks on a shared basis without the capital, maintenance, and staffing costs of on-premises message-switching computers.

Western Union also offers a low-cost CRT terminal, the Video 200, which provides text-editing capability and streamlines the preparation of messages transmitted over the InfoCom system.

Another common carrier, MCI, offers services mostly for voice users, plus some dedicated line facilities operating at data rates to 9.6 kbps. The voice channels include a variety of intercity long-distance telecommunications services to business and residential subscribers.

MCI's own microwave network runs about 8000 route-miles coast to coast. The company has tried to compete fully with AT&T in the long-distance market since 1978 when the U.S. Supreme Court refused to review a lower court order in MCI's favor. In 1980, this $376 million system served 60,000 customers in 70 major metropolitan areas.

Prior to being written off as a venture, Xerox's XTEN planned to serve some 200 metropolitan areas, assuring electronic message services, document distribution, data transport, and high-quality graphic information. The plan was to integrate satellite and city-wide radio links, featuring end-to-end digital service for interactive, batch, and memory-to-memory applications. User terminals inside office buildings were to be connected to equipment interfaces, with a transceiver and antenna on the building roof. From there the message could be beamed into the long haul network via radio link transmission to the node through a subnode, with the city nodes being the hubs of the information distribution system. Communications equipment will execute the store and forward, priority control, multiple address, delivery acknowledgment, and so on, with the outgoing messages proceeding from the city node to the satellite earth station or a microwave terrestrial link.

This is practically the structure of all other telecommunications capabilities under development, with the integration of voice, text, data, and image capabilities into one network being an overriding design characteristic. For a specific user, the desirable properties may not be exactly what a given network offers, but the network's range of facilities is such that valid technical and economic solutions are reasonably sure to be found.

Chapter 8

THE MICROWAVE LINK

INTRODUCTION

The number of telephones has doubled worldwide in the last 10 years, and the demand is expected to continue to grow at the same pace or faster. In some areas telephones are still not abundant, while in many urban regions a greater number is needed to cope with a society that is becoming busier and more complex.

The telephone system carries an enormous burden: it must not only reach every corner of the earth but also do so efficiently. This is the essence of technological breakthroughs in telephony, such as the microwave links and digital switching. Interconnecting both FDM and PCM, digital-switching approaches are fully compatible with existing networks; their flexibility makes them the nucleus of future integrated services appealing to voice, data, text, and image. Not surprisingly, a large equipment market looms ahead and is expected to reach $46 billion next year and between $68 and $70 billion dollars by 1987 (Figure 8-1), ahead of the distributed data processing market, which will probably stand at about $65 billion in 1987.

Modern communications are vital in the support of today's intricate and elaborate social and economic structure. Rising prosperity produces more people with more to say, more desire to read, and more money to

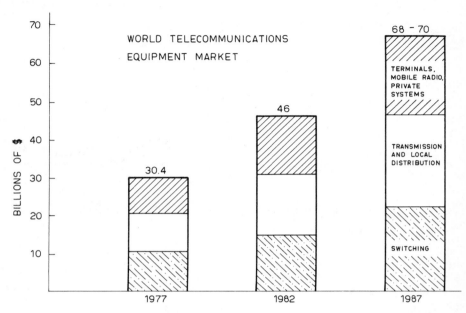

FIGURE 8-1   Statistics on the world telecommunications equipment market for 1977, 1982, and projection for 1987.

spend on communications. Efficient carrier systems are necessary to see through this sharply increased load.

Microwaves can link widely separated places in a matter of seconds; a reason why such systems have become a major trunk that can cover entire nations and crisscross the continents. Microwave communications can transmit a variety of information such as voice and TV images, as well as other newly demanded information means. Consequently, their importance is now fully recognized, and technological improvements will have a great effect on the development of all countries, from sociological to economic and business perspectives.

A microwave system can be used extensively as a trunkline to connect subscribers. With its capacity and high quality it is used to effectively provide a variety of communications services in both rural and urban areas.

Lightwave communications is another of the recent outstanding achievements in the telecommunications field. It transmits information in the form of a lightwave through a glass fiber with a capacity tens of thousands of times greater than that of a conventional carrier system. Optical fibers have already set new standards for the telecommunications systems of the future; the same is true of satellite and undersea communications.

HERTZ (FREQUENCY CYCLES PER SECOND)

10^3 10^4 10^5 10^6 10^7 10^8 10^9 10^{10} 10^{11} 10^{12} 10^{13} 10^{14} 10^{15} 10^{16} 10^{17} 10^{18} 10^{19} 10^{20} 10^{21} 10^{22} 10^{23}

(VOICE) | UNDERSEA COAXIAL CABLE | AM | SHORTWAVE BROADCASTING | FM | VHF TV | UHF | MICROWAVES | INFRARED | VISIBLE ULTRAVIOLET LIGHT | X-RAYS | GAMMA RAYS

FIGURE 8-2 The electromagnetic spectrum (in cycles per second, Hz) from voice bandwidth to gamma rays.

Why can light transmit information? Imagine the electromagnetic spectrum with radio waves for voice-level bandwidth at the left (lower frequencies) and the very high (VHF) and ultra-high frequencies (UHF) used for TV to the center right. Further on come infrared light, visible light, ultraviolet light, x-ray radiation, and gamma rays (Figure 8-2). Then notice that the amount of information that can be sent over a medium depends on the range of available frequencies in the usable band of spectrum.

In simple terms, the higher up we stand at this spectrum, the greater the range of frequencies in the band, hence the greater the *bandwidth*. Between frequencies 100 and 1000 there are 900 frequencies; between 1000 and 10,000, there are 9000; these discrete frequencies can be used to modulate carrier waves. Consequently, the more bandwidth, the more capacity we have to carry encoded data, text, image, and voice.

The problem is in developing the channels, the switching capabilities, and the equipment that can operate at the higher frequencies. For years we have tried to increase the frequencies our systems use and managed to make systems that run from 100,000 to 10 billion cycles per second (cps, hertz). Satellite systems project to broadcast at 12 to 14 billion Hz, and in the areas where lasers operate, frequencies are in the magnitude of 100 tetrahertz (100 trillion cps), given the range of light frequencies made available by the laser.

The bandwidth is very important in terms of cost effectiveness. Whichever type of line we may be considering—satellite, radio, landline, underground, or undersea—cost is a function of the bandwidth (Figure 8-3).

Dependability is another fundamental consideration. Communications serving human beings are found in surprising places. They can be seen not only spanning rugged terrain, but also high above in space and deep beneath the sea. With their unique capabilities, today's satellite and submarine communications systems are bringing remote nations closer as well as expanding existing domestic networks. We begin to see

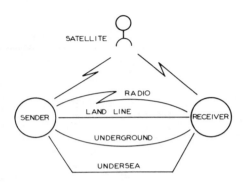

COST IS ASSOCIATED WITH
BANDWIDTH

(1) THE RANGE OF FREQUENCIES AVAIL-
ABLE FOR SIGNALING

(2) BPS

(3) THE DIFFERENCE IN HERTZ (CYCLES
PER SECOND BETWEEN THE HIGHEST
AND LOWEST FREQUENCIES OF A   **FIGURE 8-3**   Possible line connections
BAND)   from sender to receiver.

the era where technology will be providing integrated services, making advanced communications available to a greater number of people and for an ever-increasing range of reasons.

THEORY IN A NUTSHELL

The basic equation for microwaves is $\lambda \times f = c = 984,000$ ft/s ($= 3 \times 10^8$ m/s); where c is the speed of light, λ the wavelength (in meters), and f the frequency (hertz, cycles per second). Microwave energy concerns extremely high frequency and wavelengths in microwave dimension, with transmission a wave propagation phenomenon.

Microwave technology and the physical equipment it employs are divided into four main parts:

1. Generation and processing

2. Transmission

3. Measurement and control

4. Application

Dielectric media may serve as waveguides. When magnetic waves pass through a dielectric, they are slowed. The electromagnetic spectrum is broken down into radio communications, radar, and light-related emissions.

TABLE 8-1   The strata of the waveband.[1]

Waves	Length	Frequency Band	Abbreviation	Number of kHz per Meter (Change in Frequency)
Very short	10–1 m	30–300 MHz	VHF	5×10^4
Ultra short (microwaves)	1–0.1 m	0.3–3 GHz	UHF	5×10^6
Super short (microwaves)	10–1 cm	3–30 GHz	SHF	5×10^8
Extremely short (millimeter waves)	10–1 mm	30–300 GHz	EHF	5×10^{10}
Quasi-optical	1–0.1 mm	300–3000 GHz	—	5×10^{12}

[1] At the short wavelengths, the waveband is divided.

As explained in the introduction, microwaves occupy the shortwave lengths of the electromagnetic radiation region (Table 8-1). As such, they have long been vital for radar, military uses, and telecommunications. But microwaves have not yet been widely used for local communications, where telephone lines dominate.

This picture is changing, however. The costs are falling while local markets, such as cable TV and data communications, are taking off at growth rates estimated as high as 40% annually. Satellite communications open up another broad group of microwave users. As the capacity of a route increases, it is more economical to use higher data rate channels, 34, 140, and 560 Mbps being examples.

The new transmission media come at the right moment. While the television industry's mounting distribution requirements in the late 1940s stimulated the development of the first commercial microwave radio relay system, the search for higher-capacity links has made computers, communications, and image processing its gears. Except for local and short-haul facilities, the microwave relay system provides a greater volume of *bandwidth miles* than all other facilities combined.

Coaxial cable systems have been designed to meet television needs and the document distribution requirements, but radio relay carries the lure of lower first cost, even though it calls for many more technological innovations and more difficult design choices. A digital microwave radio is needed to assist cable systems in providing the digital trunks to the tandem and toll digital switches and between these switches.

Just the same, a digital microwave radio is necessary for local distribution, where a digital subscriber carrier (with or without concentration) makes economic sense. This use of digital radio will accelerate once the operating telephone companies realize that local digital switching is the answer; however:

1. Microwave technology calls for terminals that are distinctly different from those in coaxial cable systems.

Whereas coaxial systems have used single-sideband modulation, akin to amplitude modulation (AM), frequency modulation (FM) was chosen from the outset for modulating the microwave carrier by the composite voice signals transmitted over the microwave system.

2. A still more sophisticated microwave terminal is the one required by earth stations for communication satellites.

In present-day technology, a radio system is composed of a microwave transmitter/receiver and an eight-level modulator/demodulator. The latter is a major item in any radio equipment; it can provide, for example, a transmission channel for two synchronous 21.5-Mbps data streams.

Figure 8-4 presents the building blocks of a radio equipment configuration, including the transmit/receive data processors and the modulators/demodulators, which are the heart of such equipment. A direct microwave system is shown in Figure 8-5. This block diagram shows the location of filters to meet all operating objectives of radio transmission.

A basic distinction in data transmission using microwave technology is in the number of hops:

- Short haul, typically, has 1, 3, 5, and up to 10 hops.

- Long haul, by contrast, usually goes from 10 to 30 and 50 hops.

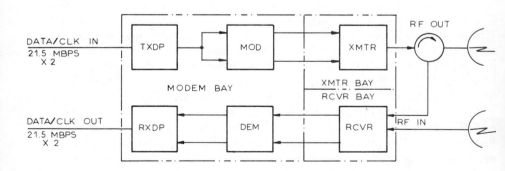

```
TX/RX DP  :  TRANSMIT/RECEIVE DATA PROCESSOR
MOD       :  8 PSK MODULATOR
DEM       :  8 PSK DEMODULATOR
XMTR      :  MICROWAVE TRANSMITTER
XCVR      :  MICROWAVE RECEIVER
```

FIGURE 8-4   The building blocks of a typical radio equipment configuration.

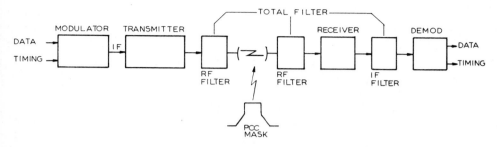

FIGURE 8-5   Block diagram of a direct microwave system.

One of the major advantages of SBS (Satellite Business Systems, as will be discussed in Chapter 12) is that through the use of relatively inexpensive earth stations, it will be possible to carry one-hop long hauls. When both data and voice are used over the same channels, basic techniques permit savings to the user.

In the future, private microwave radio systems may be employed for a wide variety of two-way, point-to-point communications services, including voice, data, telemetering, and video transmission purposes.

Separate frequency bands for operation and the allowable bandwidth associated with each system are defined in the rules and regulations governing international bandwidth agreement. Such bandwidths may be employed by public utilities and by private systems. Privately owned microwave radio systems are widely employed by oil and gas companies and by a growing class of other industrial users, including banks.

Private microwave systems may be well suited to linking a corporate headquarters facility with suburban freight terminals, data processing centers, refineries, reservation centers, chemical complexes, or branch offices. The operation of the carrier's or customer's PBX can be integrated into the radio system:

- To form a dial tandem operation
- To make each telephone station at the branch office emulate those of the main corporate headquarters

Dedicated channels for such services as data processing can be "piggybacked" over the radio using various digital or analog modulation techniques; the same is true of text and image systems.

DATA AND VOICE

The original telephone unit was projected for speech and predominantly used this way for over three-fourths of a century; however, technological advances did take place, opening new perspectives on which we can now capitalize.

When an organization needs its own data and voice communications network, the goal is to build a system that, at the lowest overall cost, will satisfactorily handle traffic from all locations and also allow traffic growth. The larger a well-designed communications network is, the lower its unit cost for the defined transaction. Up to a point, the company can benefit from the economy of scale.

However, in actual practice, design is not that simple. Understanding the cost factors can shed considerable light on the directions of communications, including the impact of tariff changes and improved technology. This is magnified for companies working worldwide with multiple pricing arrangements. For example, Pan Am has found that monthly charges for 10-mile voice channels range from $20 a circuit in South Africa to $1200 a circuit in Thailand. This same circuit can be leased in the United States for $64, in Italy for $230, and in Germany for $545.

The 100-mile voice channel goes for $148 a circuit in the United States and for $2046 a circuit in France and Germany. Prices for the 300-mile voice channel begin at $270 a circuit in the United States and go all the way up to $2590 a circuit in Sweden. And most telcos around the globe think of revising their tariff structure to discourage users from setting up private networks, the idea being that tariffs should induce shared use with privacy guaranteed through closed user group (CUG) software.

Higher tariffs are an inducement for companies to look into the multiple use of the lines they lease. Current technology permits a rate of 64 kbps per voice channel, and much more can be done with the capacity than straight voice transmission. Data and voice signal transmission are examples, although video is a very large consumer of available channel capacity.

Defining a wideband as any frequency band larger than the one we need at this very moment, and supposing we have both data and voice requirements to handle, transmission can take place with data-above-voice (DAV) or data-under-voice (DUV). A DAV system for 1.5 Mbps is shown in Figure 8-6; a DUV system for the same bandwidth in Figure 8-7. The two of them can be contrasted to a block diagram of a dedicated data system (same bandwidth) shown in Figure 8-8.

Other possibilities in frequency-division multiplexing are data-in-voice (DIV) and data above video (DAVID). Such solutions divide the channels available for transmission and allocate them among the processes competing for the resource. In every case, the objective is great efficiency in using the available facilities.

The combining of many signals is called "multiplexing." It can be accomplished in either of two principal ways:

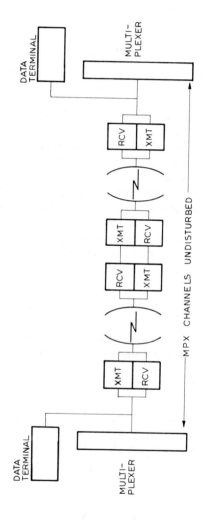

FIGURE 8-6 Block diagram of an FDM transmission system with data-above-voice (DAV).

- Frequency division
- Time division

In frequency-division multiplexing (FDM) different signals are assigned to different frequency bands to which they are usually translated by single-sideband modulation. They are then all transmitted simultaneously; at the receiving terminal, the signals are individually demodulated.

Time-division multiplexing (TDM) resembles the allocation of slots. Different signals fit in each time slot (Figure 8-9) through multiplexing. Typically, such signals can be voice, data, text, and video image, according to the transmission requirements with which we are faced. Once again, both technical and financial considerations should guide our choice. Some comparative data for FDM and TDM (among other solutions) are given in Figure 8-10.

One of the significant products in a multiplexing environment is the statistical or intelligent multiplexor to which a brief reference was made in the introductory chapters. The basic attraction of the new TDMs is the improvement in network utilization which they offer, accomplishing a greater efficiency by taking advantage of network statistics.

If we regard a circuit as the path between, say, a processing unit and a terminal, we can view its use from three statistical aspects:

- Circuit usage
- Data activity
- Code or language usage

In terms of the first reference, the dimensions of the connection periods vary from occasional to dedicated employment. Data activity refers to

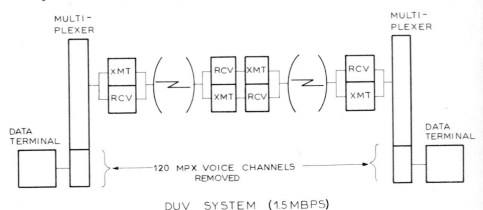

DUV SYSTEM (1.5 MBPS)

FIGURE 8-7   Block diagram of an FDM transmission system with data-under-voice (DUV).

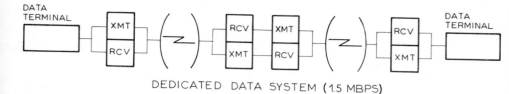

DEDICATED DATA SYSTEM (1.5 MBPS)

FIGURE 8-8 Block diagram of a dedicated data system, at about the same bandwidth as that in Figs. 8-6 and 8-7.

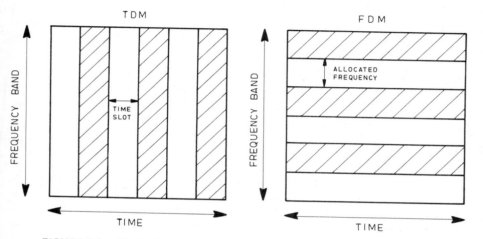

FIGURE 8-9 Block diagram of time division multiplexing (TDM) and frequency division multiplexing (FDM) plotted against a time and frequency band.

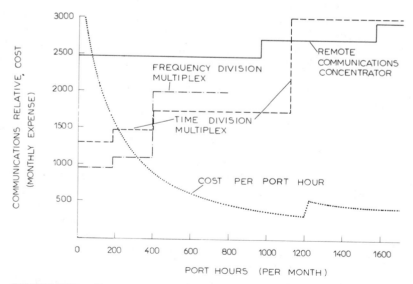

FIGURE 8-10 Comparative data for FDM, TDM, and remote communications concentrator (relative cost versus port hours per month).

the periodicities of data transmission on a connected circuit. Finally, in terms of language it is possible to account for the code being employed (Morse, English language letters, and so on). The term statistical multiplexing has come to mean those techniques that are based on the statistics of data activity.

An important, original choice is, of course, that of analog or digital transmission. With digital transmission, multiplexing becomes a straightforward process:

1. Digital streams resemble one another whether for voice, text, data, or video signals.

2. Once the signal exists in digital form the technical requirements reduce themselves to the handling of error rates.

3. There is no technological problem in data streams involving voice, data, image, and facsimile.

4. The more homogeneously digital a stream is, the easier it is to multiplex the pulses.

The problem is economic justification. Picturephone signals for integrity transmission, for example, will require a bit rate nearly 10 times higher than the current 64 kbps needed to reproduce the voice over wire in hifi (good for analog, but not for digital transmission). The bit ratio between common quality voice and video telephone is 1:300, but other factors also affect speed considerations.

Several technical questions must be solved: for example, the "call setup time," and the fact that data may be transmitted for only a few seconds, *not* typical of voice communications. But these are not difficulties from a transmission standpoint. There are two types of integration that can be done in such a network:

1. A physical integration of both digital transmission and digital switching

2. A service integration, in which voice and data are handled by the same transmission switching equipment

The second type of integration underlines the economic benefit because it eliminates the need for completely separate networks for the two services. A highly efficient digital carrier is the next major building block for the integrated digital service.

Among the advantages we observe are (1) quality service (the reason is that digital data are already structured); (2) the fact that digital data allow store and forward capabilities and, therefore, do not require simultaneously online send/receive, which greatly simplifies the transmission and distribution problem; and (3) feasibility in implementing flexible routing while the analog solution calls for rigid connections.

Digital solutions allow many-to-one relations because of the storage capability made available. The form of the dialogue procedures has also changed, and the transmission itself is greatly revamped by statistically multiplexing data and encoded voice in a transparent and intelligent network.

Typically, such a network is computer-supported. A new software implementation of a protocol for encoded voice transmission takes advantage of the 60 to 65% idle time in conversations (one way), so that only active periods in speech need be transmitted. This is not possible with present frame synchronous vocoders (voice encoders) which transmit continuously.

THE NEW VOICE COMMUNICATIONS

During the coming years, we will experience important progress in both online digital computer communications and digital voice communications, benefiting from both real-time and store and forward solutions. The growth in interest is both an aftermath of technology and a fact due to the increasing importance of geographically distributed computation, the expanding need for resource sharing, and the requirement for securing high-quality voice communication.

The progress in digital voice communications, aimed primarily at low data rate and high-fidelity speech communications, has led to various vocoding techniques. The progress in real-time computation, aimed mainly at low delays, high bandwidth, and high cost effectiveness, leads to technologies such as packet-switching, packet radio, satellite communications, and intercomputer protocols.

Digital voice techniques and corresponding network architectures have received considerable attention. The benefits range from:

• Noise and crosstalk immunity, to

• Data and computer capability, security, bandwidth conversation

Voice analysis/synthesis devices, digital telephone sets, and computer answerback systems are already available, but voice codification must be coupled with a network concept which allows real time sharing of resources and guarantees availability.

The best solution appears to be a packet-switched arrangement. Among the key parameters to be quantified are blocking probability, end-to-end packet delay, interpacket mean (and variance) gap time at delivery, and end-to-end packet loss probability. Parallel to these developments, packet radio technology is emerging as an efficient, secure way to transmit computer data between remote sites via radio links.

Packet radio technology has numerous attractive characteristics:

low bit error rate, security, capability of handling "bursty" traffic, efficient spectrum utilization, failsoft operation, support for mobile users, capabilities for internetworking, remote access of distributed databases, and fast deployment. A typical packet radio network is a store and forward packet-switching system consisting of nodes that share a single radio communications channel. As such, it assures:

1. A netted array of possible redundant repeaters for area coverage as well as reliability

2. Design techniques which permit repeater shutdown except when processing packets

3. Network protocols to locate and label repeaters, determine packet routing, allocate resources, and permit remote debugging

4. Distributed control of network management functions among multiple stations and repeaters

A packet-based (generally referred to as "packetized") voice communications model assumes a speaker talking into the source vocoder (SV), which converts the speech waveform into a bit stream. The bit stream is packetized by the transmitting process and given to the network. At the destination, the network delivers the packets to the receiving process, which feeds bits into the destination vocoder (DV). The DV synthesizes a waveform which (hopefully) sounds like the original one.

The source vocoder efficiently converts the speech waveform into a bit stream if it uses as low a data rate (bits/second) as possible while maintaining a certain fidelity or (dually) achieving the highest quality for a given data rate. There are two main types of vocoding methods:

• Those which synthesize waveforms that *look like* the original ones

• Those which synthesize waveforms that *sound like* the original ones

Technically speaking, there are three major trade-offs: compression versus computation, compression versus fidelity, and compression versus robustness. The choice of the ideal balance is still a research topic.

The reproduction quality of the total end-to-end communications depends on the acoustic quality of the speech reproduction (fidelity), the total delay between the generation of the original waveform and its reproduction at the destination, and the continuity (smoothness) of the reproduction. Although the quality function is still not well defined, it is clear that the higher the fidelity, the smaller the total delay, and the more continuous the communication, the better the total communication. Furthermore, digital speech representation should be compressed for efficiency of communication.

In a network environment, as the total network load decreases we might wish to use a higher data rate to improve the fidelity. Similarly, when the network load increases, the data rate should be decreased to minimize delays at the cost of fidelity. These dynamic data rate changes can be accomplished by changing either the packet size or the vocoding period, but due to hardware inflexibilities (for instance, the use of analog filters), it is usually more desirable to change the packet size.

In systems with a large vocoding period relative to the sampling period, it might be as easy to change the vocoding period as it is to change the packet size. In variable systems, both factors may change: the packet size according to the complexity of the acoustic features (of their differences) in the waveform (long vowels require less description than most other sounds), the vocoding period according to the variability of these features.

A number of benefits may result from superimposing voice digitization and packet radio technologies to obtain an integrated system: for example, the efficiency, flexibility, and security obtainable by voice digitization, together with the mobile and rapid deployment features of the packet radio system, in conjunction with bandwidth conservation.

The feasibility of a packet voice on packet radio (PVPR) communications system has been demonstrated by preliminary empirical studies sponsored by Arpa. Such studies proved that:

- A distributed routing algorithm must be employed rather than a centralized one to bound the end-to-end delay to an acceptable level.

- Small packet lengths (200 bits) are needed for vocoder-like digitization rates.

- Sufficient repeater density must be provided so that the average distance between two communicating parties is about two to three hops.

This requires a network with radius of no more than two to three hops. But technology is evolving and the outcome of the early phases, already satisfactory in itself, will be outpaced tomorrow.

Chapter 9
OPTICAL FIBER
COMMUNICATIONS

INTRODUCTION

Bell Laboratories invented the laser. Laser devices are small, emit single-color infrared light, and use little energy. As a result, engineers began to think of them as signal sources in lightwave communications. If workable solutions are developed, lasers allow much greater amounts of coded information per unit of spectrum than do conventional media.

With this background, since the 1960s lasers have grown from laboratory devices with short life spans to very effective light sources with life spans of several years. At the same time, simpler and cheaper units, the light-emitting diodes (LEDs), also emerged as a possible light source, while advances in semiconductor technology made it feasible to detect tiny light signals. These detectors, the photodiodes, translate light signals back into the coded information.

The interest in this development centered on the missing link: a system for transmitting the light signals. This medium, the optical fiber, made principally of silica, has been emerging as a competitor to copper in many communications applications. It is one of the most abundant materials on earth.

When suitably engineered, optical fiber cables may be used in a variety of applications where twisted copper wire pairs, coaxial cable, and metallic waveguides are now used to transmit information. Applications range from short data links and equipment interconnections within a building, to long telecommunications trunk circuits connecting switching offices within a city or between cities, and to the medical sciences.

The small size of the individual fiber, the allowable bending radius of the fiber cable, the large information capacity, the flexibility of system growth, the freedom from electromagnetic interference, the immunity from ground-loop problems, and the potential economy are some of the features that make optical fiber systems more attractive than copper. Advances in research on optical fibers and cables in the past few years have been accompanied by similar progress in the development of optical devices and components, and optical repeater techniques.

Signal attenuation in fibers as low as a fraction of a decibel per kilometer and pulse dispersion as small as a few hundred picoseconds per kilometer have been reported. Various techniques for cabling, splicing, and connecting have been developed, but the biggest advantage of fiber optic cable is its wide bandwidth. Because potential information-carrying capacity increases directly with frequency, the availability of laser-driven fiber links provides the potential for transmitting data at speeds up to 10^{14} bps.

Lasers, emitting nearly a monochromatic beam of intense light in the visible or infrared region, have opened up a portion of the electromagnetic spectrum where frequencies are 10,000 times higher than the upper range of the radio-frequency band. Multimode optical fibers assure a transmission medium for harnessing this vast potential of lasers, thus achieving transmitting rates greater than 1 to 2 GHz.

Wide-bandwidth performance not only advances the opportunity for high-speed transfer, but also makes it feasible to multiplex numerous lower-speed channels, overwhelming twisted pairs and exceeding the transmission capability of the coaxial cable. Optical fibers are essentially immune to electromagnetic or radio interference from sources such as electric motors, relays, power cables, or other inductive fields, and radio or radar transmission sources. This creates a reduction in bit error rates.

The bit error rate in fiber optic data connections exceeds 10^{-9} versus a 10^{-6} bit error rate for metallic connections. Because of higher noise immunity, a system designed with fiber cable does not require as many error checks as the wire system, thus increasing overall performance. Among the current industry leaders in fiber optics we distinguish Bell Labs, Corning Glass, ITT, IBM, Siemens, Fujitsu, Hitachi, and NEC.

GUIDED LIGHT SYSTEMS

The single optical fiber is thinner than hair, and the fiber cable is just a half inch in diameter (12 mm). Each fiber (5 mils in diameter) offers a bandwidth over several thousand digitized conversations. Particularly underlining the new technology are such factors as:

- Costs
- Bandwidth
- Noise immunity
- Size
- Weight

Because of bandwidth and error characteristics, fiber optic transmission systems offer powerful advantages over conventional coaxial cable and metallic wire links. Benefits include lower weight, lack of crosstalk, complete immunity to inductive interference, and the potential ability to deliver signals at lower costs.

The technology behind the elements of fiber optic transmission systems has matured rapidly, padded by important developments in optical sources, cables, detectors, couplers, modulators, repeaters, and modems. If line drivers cannot do the job, next to be considered are limited distance modems (LDMs).

The LDMs are simpler and less expensive versions of conventional telephone modems. Their cost advantage compared with a classical modem increases with the data rate because the major functions performed by conventional modems can be eliminated or relaxed. Although at low data rates the cost difference is not great, at high speed the difference can be significant.

Because of its end-use characteristics, the optical fiber may indeed be the land-based answer to the projected competition from satellite carriers. Although in the next 10 years satellite carriers will divert substantial traffic from the network in the sky, this traffic may return to the earth through optical fiber transmission systems.

If the optical fiber promise materializes, we will have to think very differently about networks. The limited transmission resource allocation problems will be over, and different cost optimization techniques will be required to face the new opportunities. In five to ten years, technologies that look very efficient today may be very inefficient compared with the next level of accomplishment.

The characterization of an optical fiber as a transmission medium involves determination of the usual wave-propagation properties:

1. Loss

2. Delay

3. Delay distortion

For cable development, data on mechanical requirements and environmental susceptibility are needed in addition to the A, B, C's of the fiber itself.

Pristine fibers are stronger than steel yet very flexible (10 cm bend radius). Fiber technology itself is very old; the early Egyptians produced fibers through a pulling process. The current manufacturing technology is a chemical vapor deposition (CVD) process with required temperature in the range 700 to 800°C.

A most significant feature of fiber fabrication is its simplicity. A common technique is first to manufacture a large-scale version of the desired fiber, which is known as a "preform." Such a preform might be 2 cm in diameter and 50 cm long. It can be fed into an electric furnace and a fiber drawn down without any contacting die. This may be accomplished in a single pass at a fiber-meter per second.

The 2 by 50 cm preform would produce more than 10 km of fiber actually fabricated in one piece. It is essential to put a protective coating on the fiber as it is initially drawn to protect the surface from mechanical abrasion and atmospheric constituents which would otherwise deteriorate its strength.

Like wire cables, high-wave fiber cables require connections. One of the major differences between wire cables and optical fiber cables is the technology and expected field practice for joining two cable sections. Figure 9-1 presents the layered solution to the construction of an optical fiber cable. The connector, the stranded core, and the five protective layers can be distinguished easily. Figure 9-2 shows a cut of the cable. Two alternative construction approaches are outlined.

To protect the glass fiber from possible breakage due to rough handling and potential degradation due to harmful environmental effects,

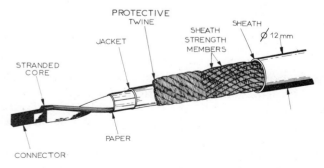

FIGURE 9-1   A layered solution to the construction of an optical fiber cable.

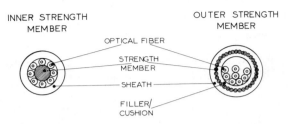

FIGURE 9-2   Two alternative construction approaches to optical fibers (inner and outer strength member).

it must be suitably packaged in a cable. The cabled fibers must also be guarded from additional radiation loss caused by excessive curvature during usage and by microbends, minute but sharp curvatures, introduced during the manufacturing process.

Although fiber manufacturing seems to be under control, the technology of optical communications links is still in evolution. Some of the work to be done in fiber optics includes the development of a reliable laser source, fast detectors, and new coupling techniques.

Fiber connectors are needed wherever detachable connections must be made: between optical transmitters/receivers and at fiber cable ends. They must be:

- Rugged

- Abrasion resistant

- Able to maintain the stringent alignment tolerance required for low loss after many repeated connections and disconnections

The methodology that will make broad use of fiber optics possible encompasses several basic elements:

- After appropriate interface electronics, a generator converts electric signals into varying light.

- At the receiving end, a photodetector performs the reverse conversion.

- At the center of the system is the fiber optic cable itself (Figure 9-3).

The modulation format for optical fiber transmission can be either analog or digital. The analog modulation of a light source (particularly light-emitting diodes) has the appeal of simplicity and economy, but the large signal-to-noise ratios required of analog systems limit their use to relatively low-bandwidth, short-distance applications.

Pulse-position modulation takes advantage of bandwidth expansion to achieve improved noise immunity and therefore wider repeater

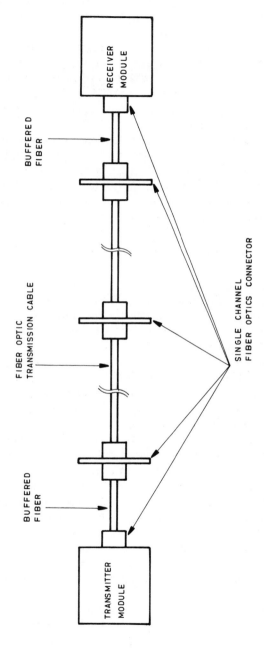

FIGURE 9-3 Transmission systems configuration for fiber optics with buffered fibers and single channel fiber optics connections.

spacing. Digital modulation is highly immune to noise and ideally suited to fiber transmission where the medium has a larger bandwidth.

The fact that lightwave systems are generally most efficiently used for digital transmission fits with the telco requirements, as digital telephone (T1) transmission systems are extensively deployed in metropolitan areas. A lightwave system at the T1 rate (corresponding to 24 digital voice channels) could interface directly with existing T1 systems. Similarly, lightwave approaches at higher speeds could use existing digital multiplexers to interface with T1.

Digital repeaters for optical fiber transmission systems are very similar to those for conventional copper wire systems. Optical repeater system experiments conducted during the last few years have produced results covering a wide range of data rates. Current repeater research interest is directed toward pushing the frontiers of high-capacity systems and broadening the base of various areas of applications.

THE APPLICATIONS VIEWPOINT

Fibers have the potential of being applied wherever metallic cable is used: undersea cable, long-haul buried cable, underground short-haul trunking, local loops, customer premise wiring, and wiring within equipment. Metropolitan trunking is a use particularly suited for the early implementation of fiber optics.

Two optical fiber cable transmission properties that will strongly influence economic viability in the whole range of applications are optical loss and signal distortion (pulse spreading or dispersion). *Low* loss means long repeater spans, and *small* dispersion implies large transmission bandwidth over long distances. These properties will also influence the selection of system components and configurations.

Fiber optic cables provide complete isolation between transmitters and receivers, thus eliminating the requirement of a common ground. Their dielectric quality presents an advantage for systems installed in dangerous gas atmospheres: they can traverse hazardous areas without igniting volatile fumes. Just the same, they assure a reasonable degree of network security against unauthorized access to confidential information. Although not absolutely fail-safe for unauthorized tapping, fiber optic cable does not radiate signals, and it offers substantial size and weight advantages over metallic cables having an equivalent bandwidth.

Still, for many short-distance applications, such as data links within a building where fiber runs are likely to be less than a few hundred meters, loss and dispersion need not be very low. Multiple fibers satisfying these moderate requirements are now available commercially.

Examined under a different perspective, the large information-carrying capacity of optical fibers is well matched to the large circuit cross sections in metropolitan areas. The small fiberguide cable answers the need for efficient use of underground metropolitan conduit. The low loss makes it possible to span interoffice distances in metropolitan areas without necessitating intermediate repeaters in manholes.

If no repeaters are to be used in manholes between central offices, loss should be less than 10 dB/km and pulse spreading below a few nanoseconds per kilometer. Requirements are most stringent for long-haul, high-capacity intercity systems. There, loss must not exceed a few decibels per kilometer and pulse spreading must fall below 1 ns/km.

The low dispersion requirement dictates the tightly controlled graded index profile or single-mode fibers. Pulse spreading as low as 200 ps/km has been observed in experimental graded index multimode fibers. Single-mode fibers with 0.75 dB/km loss have also been produced.

One of the best parameters for comparing the performance capability of twisted pair, coaxial cable, and fiber optic cables is the product or bandwidth times distance:

- For common twisted pairs, the bandwidth distance parameter is 1 MHz/km.
- For coaxial cable, 20 MHz/km.
- For fiber optic cable, 400 MHz/km.

A cost-performance factor can be computed by taking the average cable costs and then dividing the bandwidth distance parameter into the cost per kilometer. This gives interesting relationships between the approximate cost per megahertz of bandwidth for a kilometer:

- Twisted pair, $300/MHz/km
- Coaxial cable, $450/MHz/km
- Fiber optic cable, $10/MHz/km

Both financial and technical reasons account for the increasing acceptance of optical fibers in communications, replacing the coax in linking peripheral to computers. Current technology allows up to 130 Mbps over 7 km without repeater (Figure 9-4). A New York City corporation links 1200 CRTs to a series of central computers via optical fibers.

Economic perspectives are enhanced by the fact that many metropolitan areas may not even require repeater stations. The Chicago experiment performed in 1977 by AT&T offered: picturephone meeting service; trunk service; data services; common telephone facilities; and multiplexing voice, data, and image.

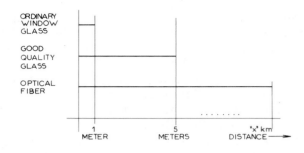

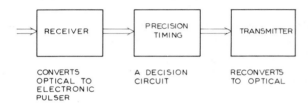

FIGURE 9-4   Light transmission in different types of glass made of silicon, and a block diagram of the repeater/regenerator station for fiber optics.

The long awaited picturephone meeting service of AT&T is becoming operational. Initially, this bidirectional color video teleconferencing will connect New York and Washington, D.C. But prior to the end of 1982 it should be connecting 16 cities. The interconnection of 42 cities is the goal prior to the end of 1983.

Not only New York, Washington, Philadelphia, San Francisco, Los Angeles, Boston, Chicago, Pittsburgh, but also Detroit, Dallas, Houston, Cincinnati, Cleveland, Columbus, and Atlanta will be among the cities to be served through picturephone this year. Let's not forget that experimentation time has taken the better part of two decades. The idea, along with an early experimental system, was launched in 1964 during the New York World's Fair.

In the Bell System's test installation in Chicago, each ½-inch cable, with 24 fibers in it, could carry over 8000 simultaneous phone conversations. One single-fiber cable has carried over 30,000 telephone messages over a distance of 6 miles, and AT&T announced an experimental fiber optic system that operates at 500 Mbps. Researchers expect 1 Gbps is possible, making the optical fiber solutions five-and-a-half orders of magnitude larger in capacity than twisted copper cables. (Voice-grade lines operate in the range 1.2 to 2.4 kbps.)

Repeaters will be needed long-distance, but metropolitan areas do not present acute problems. In Manhattan, 60% of cables run less than 2 miles, 80% less than 3 miles, and all run less than 5 miles. In Boston, 20% run less than ½ mile, and all cables run less than 7 miles.

BASIC TECHNICAL CHARACTERISTICS

Since fiber optics transmit information on photons, and not electrons, they are immune to most problems associated with electrical and radio systems operating in an environment filled with radio waves and electrical disturbances. Neither do they require links to ground in order to work. To appreciate the importance of this reference, let us recall that much of the cost in implementing communications networks arises from attempts to minimize crosstalk and interference. Fiber optics thus offer solutions that help reduce cost while providing higher bandwidth, immunity from electrical interference from other communication links and from outside disturbances (such as electrical storms) as well, and the ability to implement digital communications.

The basic components include the transmitting/receiving devices and the optical fiber waveguides and cable which assure the transmission path: different waveguides are of different purity and light-carrying capability. As transmitting devices, the lasers emit pure light. For single-mode fibers, a single-mode laser source is usually used through a lens system to excite the input end of a fiber with a spot size that matches the field distribution of the normal mode of the fiber.

Laser reliability now stands upward of 10^7 hours MTBF (mean time between failure), up from 10^4 hours in 1975. At the 50-Mbps range, lasers are fast enough. There is no problem in getting light sources, but there is a problem at x times the 100 Mbps range: reliability, bandwidth, and power happen to be competing ends, and an optimization is necessary.

Presently, LEDs and injection lasers are the most suitable sources for optical fiber telecommunications. Both are directly modulated by an input current that injects carriers across the p-n junction of a semiconductor diode.

In the case of LED, the recombination radiation is emitted incoherently from the junction. With lasers, mirrors formed by cleaving the end faces of the semiconductor crystal result in an optical cavity in which stimulated emission occurs over a narrower range of spatial and temporal frequency.

The principal advantage of the LED over the laser is that of a simpler, more economical, and more reliable structure. It produces a nearly linear relationship between current input and light output. Hence the LED is suitable for analog intensity modulation. At larger bandwidths, injection lasers tend to be preferred.

The sources of dispersion are:

1. *The optical source "rise time"*

Turning on and off can be done just so fast. Basically, an electron device has constant time, but this is not the limiting problem.

2. *The detectors (electronic devices)*

These, too, are not limiting factors.

3. *The fiber itself (as a transmission medium)*

There are two principal sources of dispersion: material and modal. For the former type (index refraction of glass), a laser gives better results than a LED. The laser is a coherent light source: spatially and frequency-wise. For 15 Mbps with current fibers, dispersion will be rather material oriented. With a LED it will give poorer results: modal dispersion will come well before material dispersion. Fibers as waveguides are multimode; single-mode fibers are much smaller and more difficult to couple energy in and out.

Modulating and repeating is another key subject. The same approaches to analog and digital modulation used with conventional wire and radio systems may also be employed for transmission on optical fibers. Still, fiber optic systems have unique modulation characteristics.

Semiconductor optical sources do not permit coherent modulation and detection. Practical systems are therefore generally constrained to some variant of intensity modulation and square-law detection.

Finally, to achieve a given total communications capacity, there is a trade-off between the communication rate per fiber and the number of fibers employed. For short-haul application, it may be economically favorable to use more fibers and not exploit each fiber to its ultimate bandwidth capability. The subject of the digital repeater must also be considered:

1. The fiber repeater contains an optical source and optical detector.
2. The metallic repeater (in contrast) contains equalization for the dispersive characteristics of metal cables.
3. Flat amplification is provided in the optical repeater.
4. Equalization for modal dispersion is difficult, owing to varia-

tions between fibers, and also results in a large signal-to-noise penalty.

5. For optical fiber systems, the pulse rate is generally limited to values not requiring equalization.

Since the optical detector has a square-law characteristic, the optical repeater typically contains more internal electrical gain than a metallic repeater. It must accommodate a larger gain variation. Bipolar codes are typically used for metallic systems, but since optical systems are inherently unipolar, other means must be chosen to maintain dc balance within the repeater and to recover timing information.

Metallic repeaters are typically powered by direct current over the communications cable, but optical repeaters require external powering. These differences impose no fundamental limitation, and indeed optical fiber digital repeaters have been built at Bell Laboratories (and elsewhere) at speeds of 1.5, 3, 6, 45, and 274 Mbps.

Chapter 10
IMPLEMENTING
AN OPTICAL TECHNOLOGY

INTRODUCTION

The impact of fiber optical transmission on voice, image, text, and data communications systems will rival that of large-scale integration (LSI) on the field of electronics. The properties of optical fibers ensure that this technology will find wide application in public, private, and military/aerospace communication systems. Other areas are under development: from power generating stations, to process control instrumentation, aircraft systems, computer peripherals, and computer/communications networks.

For instance, a data-bus system using optical fiber bundles has been tested successfully in a military aircraft; trial systems with fiber cables and repeaters carrying voice and video signals have been installed in standard telephone company ducts; and while the technical feasibility got high marks in the United States, Europe, and Japan, vital economic studies have been conducted to flash out applications that are not only technically sound but also economically viable.

Among the key applications areas we can distinguish are telephony (voice, data), military (underwater systems, submarine, helicopter controls, weapons delivery, and so on), computer interconnect (CAD, CAM, machine tools, and business systems), CATV, CCTV, broadcast TV (antenna-to-studio, in-studio), instrumentation, transportation (railways and subways), automotive (microprocessor interconnect, dashboards), aircraft, power-generating stations, and security systems (residential, community, industrial plant).

Some industries may find it more profitable than others to use optical fibers: for instance, the petroleum industry (oil logging units, refineries, oil storage, pipeline monitoring), the chemical plants, printing and publishing, hospital and health services, building maintenance (energy conservation, lights, elevators, airconditioning), education (two-way TV, interclass links, and so on), and energy transfer (via very low loss fibers). Among the developing broad areas of activity we can add: electronic mail, the satellite/PBX network, and electronic marketing (picturephone-type terminals in which buyers and sellers are linked by high-speed communications links).

As distributed information networks develop in the middle to late 1980s, fiber optics will undoubtedly find widespread use, particularly in short- and medium-distance applications. Typically, such applications take place within a plant site or office. The advantages we have mentioned that fiber optics enjoys over wire—low noise susceptibility, low loss/wideband transmission, small size, and light weight—make it an attractive medium. Yet several technical problems still need to be solved.

SYSTEMS ENGINEERING AND SYSTEMS COMPONENTS

Local distribution is one of the biggest problems with fiber optics applications. With coaxial CATV, for instance, traditionally one cable provides the service, and each subscriber taps off it. With fiber optic cables, it is not yet possible to multiplex many channels together and get good performance. Industrial applications face similar challenges, although opportunities abound, especially where the nonelectrical properties of fiber optics are vital. Associated with the implementation problem is the engineering guidance necessary for designing a system as new as the waveguides, confronting the issues relative to equipping the system, and assuring its maintenance.

Systems engineering expertise should pay much more attention to technical process management. The task is to answer the question: *How* do we build the system? as distinct from, *shall* we build the system? *How* do we line up the requirements? *How* do we translate the requirements into specifications? *How* do we really test the system to assure that it fits the goals we have established? The point is that advancing technology brings us away from massive cabling requirements, yet we must think in terms of precise evaluation and support for the systems we are building.

Management will be well advised not to take the optical fiber road unless it is assured expert advice. The necessary products are available, but this is not necessarily true of the expertise on how to use them most effectively, minimizing the unknown present with every new technology.

Typically, currently available products perform both primary functions, such as local data distribution, and secondary functions, such as signal-level conversion, asynchronous/synchronous conversion, clock regeneration, and so on. Their successful applications, however, calls for experience in designing the system, interconnecting terminal units, assuring data processing security, protecting proprietary data, and performing other necessary systems functions.

Because the systems engineers knowledgeable of optical fiber-supported networks are in such short supply, most manufacturers will supply nonclassified optical fiber systems components, but they will not undertake any systems studies. For this reason (based on a project which I recently conducted), the presentation in the following pages is (unfortunately) limited to the components the manufacturers are willing to trade, and does not include the systems perspectives I intended.

The fundamental equipment in an optical fibers environment is the Data Distribution Multiplexer System:

- A first remote unit is set to provide the total of parallel data channels (six in one specific piece of equipment).

- Four channels will go to terminals near the remote unit, while the other two are assigned to terminals some distance away.

- The four terminals near the remote unit plug directly into the connectors on the remote unit.

- The interface for the remaining two units is brought through a fiber optic link to the terminals.

- The second remote unit is assigned channels 7 through 10, while the third remote unit handles channels 11 through 16.

The modular design of the Fiber Optic Data Distribution Multiplexer System simplifies the system modifications.

The central unit and the last remote unit use a dual fiber optic interface module. Each unit employs the same transmit/receive aggregate logic card which performs the parallel–serial conversion and provides the individual channel interface to a 16-position mother board. The channel cards are plugged into the appropriate mother board slots.

If the terminal assigned to a particular channel is some distance away from the remote unit, a channel card with a fiber optic interface is plugged into that channel slot on the mother board. The parallel data for that channel is then brought through a separate fiber optic link to the remote unit where it is converted back to an interface and applied to the terminal.

The number of channels assigned to a remote unit may be changed simply by removing a channel card from one unit and installing it in a different remote unit. Additional remote units may also be added to the system at any time at any location.

- A remote unit could be added between remote units 1 and 2 simply by cutting the fiber optic cable and reconnecting it to the new remote unit.

- A remote unit may be added to the end of the chain by replacing the dual fiber optic interface module at remote unit 3 with a quad module, placing the dual module in the new remote unit, and running a fiber optic cable to the new unit.

The restriction on system configurations is that the point-to-point cabling distance between the units should not exceed 2 km. Since the aggregate signal is regenerated at each unit, the overall distance from the central unit to the last unit in a four-drop system could be 8 km. However, a multidrop line is limited to a 1 km distance.

The multiplex unit supplied by one manufacturer supports full duplex link for 16 terminals from several different locations. Modular design permits configuration changes simply by exchanging cards. Full-speed intermix with both synchronous and asynchronous data is feasible.

The equipment consists of a central unit and a number of remote units located throughout the local distribution network. The central unit operates as a conventional time-division multiplexer. It provides an aggregate interface to the local distribution network through the fiber optic cable pair, and 16 separate interfaces directly to the CPU.

The multiplexer capability is 16 channels, full duplex, combined synchronous and asynchronous operation with speed intermix. The data rate is synchronous up to 19.2 kbps, asynchronous up to 9600 baud. Lower synchronous data rates must be submultiples of 19.2 kbps.

EQUIPMENT INTERFACING

Equipment interfacing is a basic challenge with all systems, as the data communications systems designer is often faced with the problem of interfacing equipments with incompatible signal levels and formats. Solutions provide a local data distribution capability over a relatively short distance—between floors or between adjacent buildings.

To achieve this purpose, a complete family of special-purpose units has been designed to eliminate costly misuse of equipment. Signal processing is projected to take advantage of higher bit rates, and care is taken to assure that data remain synchronized throughout the system. The clock distribution system consists of triplicated frequency synthesizer circuits designed to operate with up to three external reference oscillators. Some 155 frequencies are available, and a slave unit operation provides the same frequency stability and phase as the master unit. The basic unit consists of three separate frequency synthesizer cards and from 1 to 14 line-driver cards. Each card generates a total of 21 basic frequencies brought through a bus system to all line-driver cards.

Cable drivers/receivers have an operating range up to 1.6 km over standard twisted pair (multidrop). Standard units operate up to 100 kbps; special units operate from 0.5 to 6.3 Mbps. Optical fiber technology assures a valid answer to the most common problem in a local distributed system: the transfer of data between two units located at opposite ends of the network.

Cable driver/receivers are available as separate plug-in circuit cards for multichannel applications, or as single-channel, standalone units. The circuit card/connector pin configuration is identical for all driver/receiver cards, thereby eliminating any wiring modifications when changing a channel from transmit to receive or changing the type of cable driver/receiver card used.

A modem eliminator unit replaces modems in local data distribution applications. Full duplex operations (synchronous or asynchronous) are supported and the typical operating range is about 2 km. In case fiber optic modems are used, optical coupling provides a dual solution since it not only serves as a direct replacement for a modem in local data distribution applications, but also provides all features associated with a fiber optic link. This way, a terminal on the seventh floor may be connected through a fiber optic cable to a computer on the first floor.

Fiber optic interface modules are three-channel, high-speed fiber optic cable driver/receivers intended primarily for use with a local distribution multiplexer in secure applications. The basic module may also be used for any other applications requiring a medium-speed (greater than 300 kbps) fiber optic link. The principal feature of this unit is that it recovers clock from the received data for the multiplexer. The multiplexer subsequently provides three standard synchronous clock rates below the highest channel bit rate.

Another interesting unit, the fiber optic telephone, is intended primarily for applications involving secure telephone communications. In many installations of this type, the individual telephone sets are some distance away from the secured area. The fiber optic telephone provides a fiber optic link between the remote telephones and the secured area by connecting directly to any standard telephone with up to six pushbuttons (five lines plus hold).

With two fiber optic cables, this solution assures a full duplex path for one voice channel, the seven control signals from the telephone, plus a separate data channel (up to 300 bps). The effective operating range is approximately 1 km. In the secured area, the fiber optic telephone module interfaces with the console control unit.

Furthermore, fiber optic drivers have been designed to provide a simple, inexpensive method of transferring data between two points through a fiber optic cable. The major features include four different interface formats with operating data rates from dc to more than 300 kbps. One manufacturer has two models operating in the simplex mode designed to accept an external input for data regeneration.

Finally, for diagnostics purposes each multiplexer unit displays a full status for each of the channels it is processing. Additionally, each channel may be individually commanded to a local or a remote loopback mode without affecting traffic on the remaining channels. If the system is equipped with an optional interface, the loopback commands may be originated either from the central unit multiplexer or directly from the CPU. With the 232 interface, the loopback commands are originated from the multiplexer.

When we talk of equipment interfacing, we must mention a problem which makes many potential users reluctant to accept optical fibers. This problem, *multitap*, constitutes in itself the largest area of development presently required in the fiber optic industry. There is no efficient means of doing multitap with fiber optic cable.

Fiber optic tees and star couplers are available, but still very expensive. There are no other ways for converting the optical back to electrical, and electrical back to the optical, to do multitap. Besides some pitfalls in this area due to the cost, there is also the fact that if we lose one of our electrical interfaces, all information is lost from that point on. As a result, even leading manufacturers in actual applications of optical fibers have advised that though they did several fiber installations they have yet to install any multitap solutions: all their installations have been point to point.

Let's recapitulate. Multitapping on fiber optic cable is still a technical problem area. There is no easy solution due to the extreme loss that occurs at each optical tee connection. Therefore, because of the high loss and high cost of this type of product, serious fiber manufacturers do not recommend them to their customers and have not used or installed any multitap systems. Yet the same manufacturers feel that within the next few years there will be viable and cost-effective solutions to multitapping using fiber optic cable.

It follows that though the key advantages of optical fibers are low costs and enormous bandwidth, it is not to use the full bandwidth of most fiber optic cables on the market. However, by installing fiber optic cabling in a new construction during its early stages, we would never again need to change the cabling within the facility.

Using a twisted pair of coaxial cable, as our requirement increases and more data speeds are needed, the cable finds itself insufficient; this is not true with fiber optics. As our requirement gets to higher data-rate speeds, the fiber optic cable would be capable of more efficient use and would not call for replacing or upgrading in the near future.

The size of the fiber optic cable and weight are also a serious consideration. Because they are smaller and lighter in weight, fiber optic cable (in most cases) is easier to install, cutting the installation cost. On this basis, one would suggest that optical fibers be seriously considered as point-to-point solutions versus the coaxial cable multitap. In this respect, both are tested technologies worth trying.

SOME APPLICATIONS EXAMPLES

In intra- or interbuilding applications, there are two types of peripheral-to-computer interconnections:

- Close grouped
- Remote

Close-grouped peripherals include card readers, memory disks, and high-speed parallel-driven printers. These are "close-grouped" peripherals because they are normally found in a group configuration within a short distance from the mainframe.

Since close-grouped devices are usually located in an environmentally controlled room and their distance from the computer is short, cable interference problems can be solved with simple metallic cable. However, remote peripherals on a local data network can be located hundreds or thousands of kilometers away from the central processor, and such terminals can be used for applications ranging from low-speed remote job entry to interactive graphic display jobs. Data rates on such terminals are commonly high and getting higher all the time, offering a good opportunity for employing optical fiber technology.

Experience with optical fiber applications includes:

1. On-premises, local connections (factories, office buildings, apartment houses)
2. Customer subscriber loops (clear signal services)
3. Interoffice circuits (nearly remote location)

4. Intercity trunks
5. The attachment of peripherals to the mainframe
6. Research and development (R&D) on fibers to create a "flying DB"

The objective is to put the database on an 8-bit parallel fiber (byte-serial) and keep on repeating the information stored in it. This will utilize one of the main features of the fiber: wideband.

Particularly with reference to computer architecture, there are many possibilities, some arising from limitations in the current state of the art.

Modern central hardware connections are rather chaotic, and the available line between mainframes and peripherals is short. We need longer, linearly defined lines, and fiber optics can offer the solution. The coaxial cable used until now has round or flat setups, but it takes much volume and needs great redundancy to guarantee proper function.

Also, there is an order-of-magnitude difference with optical fibers. The coaxial cable is typically 2 cm in diameter; the optical fibers are 2 to 3 mm in diameter. Among the advantages of the latter we distinguish:

1. Dimension
2. Speed of propagation (coaxial: 7.5 ns/m; optical fiber: greater than coaxial by an order of magnitude)
3. Dissipated energy (coaxial: 0.5 watt; optical fiber: 0.1 watt)
4. Noise immunity (coaxial: at 30 m have shown much exposure to noise; optical fiber: present no such problem, opens new possibilities)

Furthermore, we gain on the speed of the drivers: both sender and receiver. For medium-range distances line- or cable-driving techniques are commonly employed to interconnect local terminals to a central processor. Line drivers are required because of signal distortion (which limits distance and speed) inherent in the electrical characteristics of cables. Extended-distance transmission produces pulse rounding, a condition in which the edges of a square-wave pulse are distorted because of the loss of high-frequency elements. Line drivers typically operate at speeds of up to 19.2 kbps, but special units are available that will operate at data speeds of up to 1.5 Mbps.

Optical fibers also have disadvantages: one example is present-day costs for the optical fiber installation, the end equipment, and the light source. This will change with time.

A more elusive issue is the standardization of end equipment. Today each manufacturer does as it pleases, complicating the task of interfacing with peripherals. Present peripherals are cable-oriented, and it is not easy to change them all. Also, there is a risk of dividing the peripherals into:

- Optical fiber-oriented
- Coaxial-oriented

Although nearly all computer manufacturers work in laboratories on mainframe-to-peripherals connection through standardized interfaces, there is no widely acknowledged field installation. This should not discourage us from trying. Difficulty is the mother of invention.

VIDEODISKS

The subject of videodisks may seem too far apart from that of optical fibers to be included under the same heading, yet a careful examination of the technical facts will demonstrate the opposite. Although optical fibers are used for transmission and optical disks for recording, they are both made of the same material and use lasers: the same light source.

The present videodisk players are directed at the home entertainment market and, in their present form, are not suitable for business use. However, new systems under development and emerging on the market will provide low-cost high-density digital storage with rapid random access. These devices will replace both paper and COM and make the electronic filing cabinet a reality within the integrated electronic office.

Videodisks are a major step in storage technology. They permit information to be stored and retrieved more cheaply than ever before. The storage capacity of digital products is impressive: a single disk can store approximately 10 billion bits of information. On a standard 30-cm (12-inch) disk there are more than 50,000 tracks of information organized into sectors which can be addressed at 250 ms access time.

To write or play back the disk using an optical lens system, a laser is focused on the underside. Where the surface has no pit, the light beam is reflected back through a series of mirrors onto a photodiode. When a pit passes the beam, the laser light is diffracted and gives a modulation to the output. Frame-freeze and slow-motion effects can be obtained. An important feature of optical systems is that the head does not wear because it is not in contact with the disk.

Several manufacturers are interested in the digital or professional market. "Digital Read After Write" (DRAW) disks can soon be expected on the market. In one of these systems each track of information is divided into 128 sectors, each of which can be addressed. The laser head is mounted on a radial arm and moved by a small servo-controlled linear motor. Positioning is done by counting the number of tracks passing the head, rather than by absolute positioning. Average access time is 250 ms. A 30-cm DRAW disk could store the equivalent 1 gigabyte of hard disk, seven magnetic tape reels, 100 microfiche, or 20,000 A4 page images.

Thus, in a broad sense, the optical disk market can be divided into two main classes:

1. *Consumer* (analog, digital) for video and audio HF
2. *Professional* (digital)

The professional disk market will experience several entries, two sure ones being Philips, and IBM/MCA. The audio HF compact disk has currently one standard, developed by Philips/Sony; the videodisk has three:

- *Philips* VLP—Laservision: supports stereo and still pictures

IBM/MCA, Pioneer, Grundig, and 3M follow a similar standard based on an Olympus lens mirror system.

- *JVC (Matsushita):* has less density, no still picture
- *RCA:* Selectavision (Zenith and Thorn have subscribed to this)

For professional applications, Philips projects DOR (Digital Optical Recording), a different process than Laservision. It supports read/write and has the forementioned 250 ms access time. The "1st Series" (still a prototype) of about 40 units was expected in 1982.

A "Megadoc" jukebox based on DOR is being designed for decentralized archives to handle multiple videodisks. The reader handles a page (at 4 megabits/page) and stores the contents. The monitor gives the 4-megabit resolution for reading.

Given that a storage capacity of 10^{10} bits/disk at 4 megabits/page supports about 10^4 pages. If a 50-disk battery is online, there will be 500,000 pages/jukebox, with transmission requiring a cable of 8 Mbps. Hardcopy possibility will exist at the workstation, together with an annotations faculty. Projected input is 4 megabits/page at the workstation level; for black-and-white presentation, 4 megabits is the minimum; response time is 1 s; and the target cost is $50,000 to $100,000.

Chapter 11

MULTITAP COAXIAL CABLE

INTRODUCTION

No multitap fiber optics networks are currently commercially available, although some are in the test and development stage. One of them is "Fibernet," an experimental project at Xerox, which applies Ethernet coaxial cable principles adapted to optical fiber cables. The experimental setup has carried 150 Mbps of pseudorandom data over a 500-meter distance, through a 100-port star coupler, with zero errors detected in test sequences of 200 million data bits. Distributed packet-switching experiments have also been made. (Toshiba has made two versions of an optical fiber-based local area network. Both are based on an experimental star coupler, and Toshiba finds that 50 fibers are possible per star coupler. The current lack of multitap capabilities sharply reduces interest in practical applications of this approach. NEC has also announced a local network version based on optical fibers for communicating mainframes.)

Optical fiber networks for wired buildings are not yet popular and the specialists are difficult to find. Coaxial cable is therefore in use today as the main industrial communications link in offices, factories, and industrial complexes: General Motors, Ford, American Motors, Dow Chemical, and Kellogg's provide telling examples. One such network by Interactive Systems of Ann Arbor, Michigan, a 3M subsidiary, employs a bidirectional CATV-type system in-house to carry multichannel closed-circuit TV, digital data, and voice communications.

On a single coaxial cable (Figure 11-1) can be attached mainframes, minicomputers, video terminals, nearby or remote printers, word processors, TV security, alarm systems (visual and audio), time clocks, conveyors, switches, pressure gauges, control valves, temperature sensors, thermostats, hydraulic pumps, fans, air conditioning, light control, and power consumption systems.

- Loads can be identified by building, location, and type.
- They can be assigned to time schedules, with each on/off schedule based on programmable approaches.
- The operation of a load can be based on the status of other loads or of temperature, humidity, or a host of other variables.
- Multiple terminals can be maintained, and priority assignments kept on time.

The coaxial cable which is the main carrier for these applications has been developed and implemented as an efficient medium by the

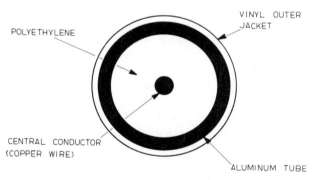

BANDWIDTH :   300 MHz

CHARACTERISTICS:

DIAMETER : 0.5 INCHES

ATTENUATION: DB PER 100 M,   20°C

    AT 5 MHz: 0.45

    AT 300 MHz: 3.9

CHARACTERISTIC IMPEDANCE: 75 OHMS

FIGURE 11-1   Cut of a coaxial cable showing the successive layers to the central copper wire.

cable TV industry. The necessary hardware is available off the shelf, and widths of 300 MHz are standard. Every major urban area has a cable TV (CATV) operator; however, coaxial components used by Ethernet and other local network architectures are not necessarily compatible with cable TV.

Two quite different ways to employ coaxial cable are "baseband" and "broadband." In the former, line voltage shifts from zero or a fixed value to some other dc value constituting the signal and, therefore, representing the data. The broadband system uses multiple high-frequency carriers, distributed throughout the 300-MHz bandwidth; such frequencies are stacked with empty intervals between in the 300-MHz bandwidth. Each carrier is modulated by the voice, data, or video information it transmits.

Such considerations account for the fact that technologies and applications are merging in communications in general, and in local networks in particular. Coming together are computers, communications, and cable television—with the result a merger of data, text, voice, and image. Companies that were never competitors before are finding themselves against one another with their competing technologies in hand.

Broadband networks are constructed from standard cable TV components, and a local data network with broadband is cabled just as if the building were being cabled for TV reception. The topology resembles a tree structure: a trunk line, branches, subbranches, and so on.

In a broadband system, there is a single-speed channel and all-users form of time-division multiple access. Favorable points of a baseband system include the cable's passivity and the totally distributed control achieved through "Carrier Sense Multiple Access with Collision Detection" (CSMA/CD). The transceivers are isolated from the cable.

In Chapter 1 we also made a distinction between baseband and broadband based on carrier capacity:

- A baseband network typically handles up to 10 Mbps
- A broadband solution supports 400 Mbps with current technology

In this sense, on a broadband installation can piggy-back several baseband protocols, with bridges provided to allow them to communicate. We will return to these notions when, in Chapter 18, we talk of local area networks (LAN).

Manhattan Cable Television is a good example of a CATV operator able to carry data traffic. The separation between the TV signals on the 300-MHz bandwidth channel is significant, and there is room in between the TV signals for digital transmission. Manhattan CATV multiplexes the bandwidth into many smaller pipelines for carrying data.

Technical problems arise from the fact that there are different kinds of coaxial TV cable. For example, the user needs an appropriate transceiver, but this can be accomplished without major problems. A greater difficulty is that to avoid restricting the use of the cable, we must have a radio-frequency modem or risk interfering with the TV signal. This can be supplied by various techniques, the main question being one of signal quality.

THE DEVELOPMENT OF TELEVISION

John Logie Baird invented television. Although none of his patents are used today, there is no doubt that he did much to stimulate interest in wide-scale transmission. Baird's actual contribution to television development must, however, be judged in relation to the work of other inventors. Characteristically, technology started bringing up the bits and pieces some 110 years before the invention.

In 1817, the Swedish scientist Berzelius discovered selenium. Twenty-two years later, in 1839, Edmund Becquerel discovered the electrochemical effects of light. These two events, seemingly unconnected, led to the 1873 announcement of the photosensitive properties of selenium, a process that set off a spate of schemes for "seeing by electricity" in England, France, and America. In the United States in 1881, Shelford Bidwell demonstrated an apparatus for transmitting silhouettes. A 24-year-old German student, Paul Nipkow, conceived, in 1884, the idea of a scanning disk, spirally perforated and rotating, which used the selenium cell.

Another 41 years passed before Baird formed Television Limited. Since 1923 he had been experimenting in television, and by 1928 he was working on color processes. His business associates tried to interest the BBC in his work, but the broadcasting authority was cool toward the idea, claiming that it would interfere with its sound services. It was eventually persuaded to allow five hours of weekly experimental transmission.

Early television viewers must have seen some odd programs. They were transmitted in two-minute segments, and only one transmitter was available, so sound and vision could not be synchronized. By 1930, synchronization was achieved and the first-ever television play was transmitted to approximately 1000 viewers, half of whom had built their own sets.

This brings under perspective the development of the cathode ray tube, and we must do a flashback. The year of the first cathode ray tube for commercial use is 1897, followed in 1907 by the concept of a system using cathode ray tubes in conjunction with mirror drums. Boris Rosing and Campbell Swinton were working on similar systems around the same time, but both systems lacked means of amplification.

World War I stimulated technical advances: the radio valve came out of the laboratory and into production and, with it, the concepts of *electronics* and *telecommunications* were born.

By 1933–1935, English, Dutch, German, and American companies were working actively on commercializing the television principle. In England, BMI and Marconi joined forces to oust Baird's television company, working on a system which had three times as many lines as Baird's and twice as many pictures. Regular BBC television broadcasts began in November 1936, with Baird and EMI systems in use. The official decision to drop Baird came in February 1937. The 240-line picture with 25 frames per second was gone forever; the 405-line with 50 frames per second became standard.

The 1937 Coronation and the use of standard sets stimulated sales. By then, there were 2000 in England, but in 1938 their price dropped 80% and by 1939 between 20,000 and 25,000 sets had been installed. Then, World War II brought European TV developments to a standstill.

In 1940, only America had regular TV transmissions, but by 1950, six countries were on the air; by 1955, 27; by 1960, 68; by 1968; 102; and today it is estimated that practically every country has regular TV transmission. Furthermore, according to figures published by the British Bureau of Television Advertising, TV became the most important source of the world news:

- In Britain it accounted for 49% of world news compared to 42% for the press.

- In Canada, 45% compared to 42% press.

- In the United States, 64% to 35%.

Simultaneously, commercial TV led to a great jump in TV viewing.

Television also brought important products, such as the coaxial cable and the technology that is needed for local networks, as well as important spin-offs on those products.

Closed-circuit television (CCTV) is a thriving market that has nothing to do with popular entertainment. Its uses fall under three headings:

- Program origination and performance analysis

- A network for the transmission of data, text, voice, and image

- Surveillance, observation, and control

Within this broad range of applications capabilities, the closed-circuit television industry is experiencing a significant growth. Part of this growth has doubtlessly been stimulated by important technical developments. For instance, the change from monochrome to color has proved significant in CCTV use for training surgeons and other medical specialists.

When today we observe these developments in action, we often forget that color television (broadcast or closed circuit) did not come in properly until the *Plumbicon* camera pickup tube was invented (by Philips) in 1963. Before this, color television cameras were so cumbersome, expensive, and insensitive that even broadcasting authorities with substantial budgets were reluctant to change.

An even more important development has been the advent of the video recorder; this facility for recording events for subsequent viewing has opened up whole new areas for CCTV application. In fact, this is one of the first mergings of computer and communications technologies. The facility for storing a program so that it can be used at any time became generally available with the introduction of ½-inch video tape recorders in 1968, and it was rendered even more convenient and attractive when the color video cassette was introduced in 1971.

Another important technical development has been the introduction of a television camera tube which is extremely sensitive to very low light levels, more sensitive than the human eye. For night surveillance, this has been a great value because the television monitor displays the picture as it would look under normal daylight conditions.

Such breakthroughs, together with the steady bulk, weight, and cost reduction of CCTV equipment, have made CCTV a practical proposition for many kinds of uses in finance, industry, commerce, sports, education, and training. An increasing number of companies are using CCTV for internal communications, ranging from conferences to research work and production techniques, while management has found a medium to present messages quickly and efficiently.

COAXIAL NETWORK

The use of coaxial cable for local networks combines the capabilities offered by the cable itself with those of the radio-frequency modems. Together, they create a new datacommunications faculty and bring into play the advantages of broadband, such as:

- Transmission over many data channels
- Ability to integrate voice with text and data
- Simultaneous handling of TV channels

Coaxial cable helps eliminate a wiring change every time an office layout or the data load varies. It is installed only once, and we can simply tap into it for future connections. As such it offers a "data pipeline," at about $0.5 per meter, for the cable only.

The complete cable system will evidently include design, installation, and commissioning, but all this is part of the job done by the

CATV industry. Part and parcel of the cable hardware are amplifiers, splitters, and taps. Included in the facility supported by the system is a high noise immunity, a large-capacity bandwidth, the low cost to buy and install, the ability to use it indoors and outdoors, the simplicity of repair (it costs less than the telephone cable), and a foolproof aspect (one cannot make the wrong connection).

Table 11-1 contrasts four alternative transport media: twisted pair, fiber optics, infrared, and coaxial. Five different factors are considered: flexibility, expandability, noise immunity, datacomm capability, and installation costs. As can be appreciated, at the current state of the art the coaxial carries the day, and twisted pair is used for nearly 98% of the datacomm installations.

Whether used for data and voice within a building or for longer hauls, the basic cable system is a one-way transmission originating at the head end, and feeding *downstream* throughout the area it serves. To transmit data, it is necessary to have an upstream as well as a downstream path to each customer location to provide a complete *duplex* circuit. The design of cable transmission is therefore such that an upstream path can be provided by the simple insertion of reverse carriage components in the amplifiers. (Either of two approaches can be taken: Wangnet and other architectures use two coaxial cables—one for forward, the other for reverse. As an alternative, forward and reverse bands can be defined on the same wire by dividing the available capacity. The 3M/Interactive Systems solution (ALAN) follows this approach. Because of CATV influence, the forward band is much broader than the reverse.)

A signal originating at one of the user locations is transmitted to the head end on a specially assigned upstream channel. At the head end (Figure 11-2) this channel is translated in frequency to the circuit destination:

1. A point-to-point circuit is established by assigning a specific subchannel to each data transmitter.

2. The capacity of this subchannel is related directly to the speed of data transmission desired.

TABLE 11-1 Alternative transport media.

	Twisted Pair	Fiber Optics	Infrared	Coaxial
Flexibility	Poor	Poor	Good	Good
Expandability	Expensive	Expensive	Reasonable	Reasonable
Noise immunity	Poor	Good	Problematic	Good
Datacomm capacity	Average	High	Average	High
Installation cost	Expensive protection	Low	Medium	Low

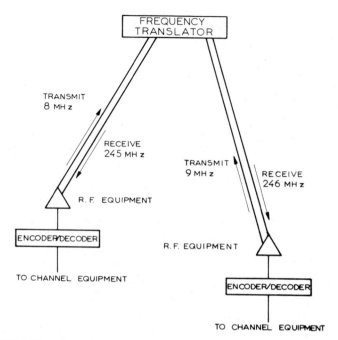

FIGURE 11-2   A cable television installation with head end and frequency bands implemented by Manhattan CATV.

3. All the various upstream subchannels are collected, and translated as a group into a downstream channel in a frequency range called the "superband."

4. This full data channel (made up of various subchannels) is then transmitted downstream.

5. At the receiving points, where data receivers are tuned to the assigned subchannels in the superband, the data are detected and then delivered to the customer in their original form.

For instance, at the Manhattan CATV application, full 6-MHz TV channels are allotted in both the upstream and downstream directions for carriage of data signals. Additional channels can be assigned as the traffic requires. Their capacity assignment exceeds 5 million bps. The digital and analog interface units are the only auxiliary equipment used to carry the data service over the cable transmission system.

After the input data are processed by the interface unit, it is on the cable as a radio-frequency signal. There is no further modulation or processing except for the single frequency translation of the entire data band at the cable system head end.

The parameters for successful CATV data transmission are: provision for point-to-point (not switched) service; sold and used strictly on a bandwidth requirement basis, independent of user's data format; and easy to operate, dependable, and readily maintainable by CATV technicians.

While using the CATV system for data transmission, the network must have the capability for full duplex operation: that is, transmitting and receiving simultaneously. As stated, this was accomplished by using one coaxial cable for transmitting and another cable for receiving: each cable is used unidirectionally. Most recent circuits, however (for example, those installed at Bankers' Trust), employ coaxial cable operation in a bidirectional mode.

The Interactive Systems' CCTV works at 300 MHz, with 5 to 30 MHz reverse transmission (control) and 50 to 300 MHz forward transmission. A distinction should evidently be made through the coaxial network's ability to carry signals and the local network architectures it can accommodate.

Even if at present such facilities are not included, accommodation can be supported through successive releases. In the case of Ethernet, for example, the current release transmits through raw digital signals, which cannot be put on CATV. However, putting radio-frequency modems between Ethernet and the cable, it is possible to restrict Ethernet to a given band. Thus "several networks" can be piggybacked on the same cable by taking advantage of the baseband facility supported by the coaxial.

Control is achieved through software, and although the local network might have been wired through twisted wire and star-type approaches (Figure 11-3), classical solutions, first, do not allow piggybacking and, second, presuppose low loads and few connections to be technically effective and economically viable. Furthermore, in classical approaches to installing a new communication, control, or information system in a plant or office building, half the cost is in the wiring.

This makes coaxial cables cost effective while also being well-established technology. (Its original use put the cost of installed cabling at $3 to $10 per foot, or $10 to $30 per meter, depending on cable size and the complexity of the installation. In the beginning, such cost impeded wide use of central systems except in highly capital intensive industries.)

Efficiency increased substantially because the new generation of coaxial cables uses single copper wire that can carry 50 TV channels or 500 data channels simultaneously, there are fewer wires to hook up, and initial installation is cheaper (less than $3 per foot or $10 per meter). In the new building where wiring must run for many floors or

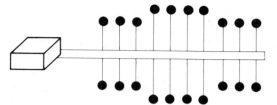

MODERN COAXIAL SYSTEM HANDLES ONLINE DEVICES
WITH A SINGLE WIDEBAND, SINGLE CONDUCTOR-CABLE
FED BY LOCAL DROP CABLES.

FUTURE EXPANSION WILL BE INEXPENSIVE.

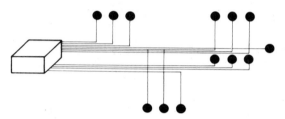

CONVENTIONAL STAR-TYPE DATA COMMUNICATIONS:
DEDICATED WIRE PAIRS OR CABLES CONNECTING
PERIPHERALS TO A CENTRAL RESOURCE.

FUTURE EXPANSION WILL COST A GREAT DEAL.

FIGURE 11-3   Modern coaxial cable loop contrasted to conventional star-type
communications served by twisted wire.

several thousand feet, the broadband network can offer dramatic savings over paired wires or individual dedicated cables. Furthermore, once the cable is in, we can add audio, video, and data channels as needed just by hooking up to the cable. We do not have to run wires back from the source to the central point of the system.

In Figure 11-4 the amplifier is the active component; all others are passive. The system accepts many FDM and TDM channels working together on the same cable but with different frequencies, using the frequency to separate the channels.

A pair of twisted wires is quite limited in the signal frequencies it can handle compared to the signals ranging in frequency all the way up to 300 MHz supported by the coaxial cable. The cable can also replace phone lines that connect a central PBX with the workstations, integrating other services (text, image) and opening the way to the technologies of the mid-1980s with video conferences and communications far more sophisticated than today's standard phone-and-switchboard approaches.

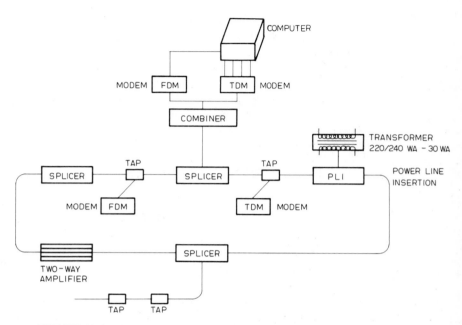

FIGURE 11-4 A systems view of a coaxial cable network with both active and passive components.

A SYSTEMS VIEW

A TV picture requires a bandwidth of 5 to 6 MHz and a 300-MHz cable can handle up to 50 TV channels simultaneously. Since a data channel requires much less bandwidth, the same cable can carry about 500 channels without any sophisticated approach. Technical solutions such as FDM (frequency-division multiplexing) and TDM (time-division multiplexing) help expand the cable's handling capacity. With TDM, each terminal or other data source is assigned its own special fraction of a second in which to transmit, although it can interrupt to use other sources' time segments when necessary. Through multiplexing, we can add up to a theoretical maximum of 10,000 terminals onto a cable.

Thus the modems used with CCTV can be of two basic types:

- FDM
- TDM

Figure 11-5 outlines the FDM solution. The computer is at the head end. Modem 1 transmits (Tx) at frequency "1" (F1) and receives (Rx) at frequency "2"; modem 2, at F3 and F4, respectively. The terminals

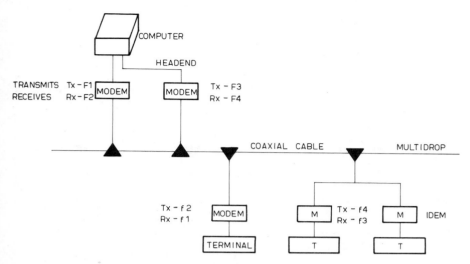

FIGURE 11-5 An FDM solution to a multidrop/multitap coaxial cable network.

connected to the cable transmit and receive at frequencies f1, f2, and f4, f3. It is possible to multiplex a number of terminals, but these terminals should be kept close to the multiplexer.

As most terminals do not work on high data rates, we can divide channel capacity by time. In TDM solution:

- Single-frequency channels are used.
- The head end can communicate with up to 250 addressable modems.
- Each modem is addressed in a time slot.
- Returned data do not require polling (returning automatically).

Figure 11-6 shows a typical TDM arrangement. At fast cable database (100 kbps) one byte is received from 250 modems in 32.5 ms. The organization is simple to expand, and it is possible to mix TDM and FDM (as most people do) if the network system has a built-in checking procedure. (Different channels will be dedicated to TDM and to FDM.)

Some architectures allow more than one computer to be used at the head end with the operator of the terminal being in a position to actuate selection of the mainframe. In its current release, Interactive Systems, for example, supports several mainframes at the head end, and the system can work either in FDM or TDM (Figure 11-7). Table 11-2 outlines some examples of European CCTV applications.

Thus, viewed as a communications system, the CCTV/CATV potential far exceeds the transmission requirements of the channels devoted

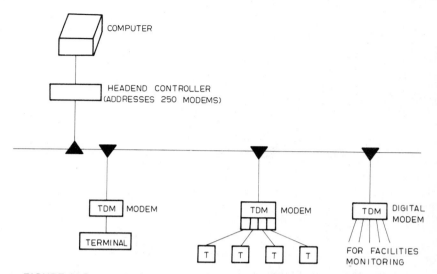

FIGURE 11-6   A TDM arrangement with head end controller and modems serving one terminal, more terminals, and facilities monitoring devices.

to entertainment television or video control. Let's look at the advantages it presents in the whole communications field.

1. *Serviceability:* Inherent in the CATV scheme is the fact that all signals, whether television, radio, or data, are transmitted on the same coaxial cable.

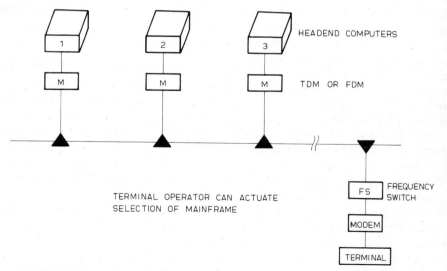

FIGURE 11-7   A systems approach able to handle more than one head end computer with frequency-switch capabilities.

TABLE 11-2 European CCTV applications.

Company	Technique	Number of Modems	Systems
Leyland Vehicles	FDM	46	1
Ford	FDM/TDM	200	6
Leyland Cars	TDM	180	4
Talbot Cars	FDM	11	1
Landrover	FDM/TDM	138	5
British Airways	FDM	140	1
Philips, Eindhoven	FDM/TDM	8	1
Philips, Leuvan	TDM	55	1

This adds to the reliability and serviceability of the system. Monitoring equipment located in the head end will automatically switch between these two trunk cables in the event of signal degradation, providing complete redundancy on signal paths.

2. *Repairability:* The restart and restore time in the event of service outages on the CATV system is generally better than on the equivalent telephone circuit.

Since all signals on CATV are transmitted through a single coaxial cable, disasters such as a cut cable can be repaired with only one connection. In classical telephone cables, on the other hand, as many as 2000 twisted pairs need splicing when damage occurs. The inherent ability of a CATV technician to pinpoint network problems quickly and *without* the user's intervention, is a marked advantage for the system.

The performance of the cable system is inherently better, and the burden falls on the cable operator to keep it this way. The basic cable system that carries the data services is maintained by a crew of experienced technicians using the latest equipment and techniques. Personnel on a 24-hour duty schedule are equipped with the necessary vehicles and instrumentation to respond quickly.

In addition, some systems, such as the one supported by Manhattan CATV, are fitted with a network of automatic performance monitors which feed a continuing stream of maintenance information to the control center. This allows early sensing of system degradation and facilitates preventive maintenance.

Catastrophic trunk failures, such as a cable being broken by construction machinery, are infrequent but inevitable. In such situations the cable system has one overwhelming advantage: there is only one cable (not hundreds of wires) to repair.

In the case of CATV, throughout the system certain trunk lines are used to feed large areas of the city. For instance, with Manhattan

CATV, the trunk feeding Manhattan's midtown and the downtown financial district is being backed up with redundant cable on different routes and with automatic switchover equipment. Should there be a failure in the main trunk, the second trunk will automatically take over with the loss of only milliseconds of continuity.

3. *Cost:* The excellence of the medium allows less terminal equipment complexity for services such as high-speed data transmission.

It has been reported that Bankers' Trust Company has realized savings of 25 to 50% over the cost of equivalent telephone company wideband channels. Success stories understandably expand the applications perspectives. Figure 11-8 presents current forecasts for expanding baseband and wideband coaxial systems into 1990. Optical fibers are not expected to overtake wideband until 1986 and baseband coaxial after 1988.

4. *Flexibility:* Since all transmissions are sent on one cable, once a building is wired, all new requirements can be accommodated with minimal additional cabling.

Increased throughput accompanies flexibility. The CATV hardware employs a highly effective coding technique and a unique filtering method resulting in a high-performance *data highway*. The high signal-to-noise ratio on the CATV network makes it an inherently clean medium for the transmission of computer data. Tests at Bankers' Trust indicate that

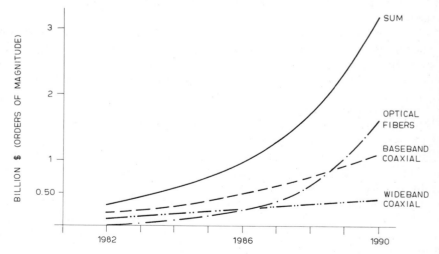

FIGURE 11-8   Current forecasts for expanding wideband and baseband coaxial cables and optical fiber systems in 1990.

bit error rates are two orders of magnitude better than can be expected on conventional analog data lines.

AN APPLICATIONS PROFILE

In the case of the Manhattan CATV applications, a signal originating at a customer's location is transmitted to the head end on a specially assigned upstream channel. A point-to-point data circuit is established by assigning a specific subchannel to each data transmitter. The capacity of the subchannel is directly related to the speed of data transmission desired. The upstream subchannels are collected at the head end and converted into a downstream channel in the superband frequency range.

At the receiving point, the data are detected and delivered to customers in their original form. The capacity can exceed 5 million bps. This facility did not escape the attention of major New York banks, and these financial institutions have adopted this service for their internal communications:

- Citibank
- Chase Manhattan
- Manufacturers Hanover Trust
- Morgan Guaranty
- Bankers' Trust
- American Express

The coaxial cable's clear signals can be offered point to point, polled, and full duplex from 4.8 to 50 kbps. (The computer-to-computer communications network was brought to the threshold of voice communications, multiplexing voice, and data, but AT&T objected and the project probably will not continue.) Commercial bankers in the Manhattan area, however, have seen the possible advantages and have taken the lead in using cable television channels to transmit data between operations centers.

Bankers' Trust instituted a pilot test of the cable line between the bank's former operations center at 16 Wall Street and its BankAmericard division at 1775 Broadway. During the four-month test period there was only one breakdown of the cable, and that was repaired in 75 minutes. An equivalent telephone line breakdown (3000 twisted pairs of copper wires) could take three to four days to fix.

Cost is another major issue. Bankers' Trust has been paying $600 for each of its two cable TV links since 1974, but the price of each telephone company link was $800 when the changeover to cable was installed in 1974. (Manhattan Cable Television may be forced by a state regulator to raise its prices. The New York State Public Service Commission has asked Manhattan CATV to show why it should not be considered a common carrier like the state-regulated New York Telephone.)

Potentially, cable links can be implemented for intracity point-to-point debits and check verification systems. Because they have full duplex properties, allowing messages to be received and transmitted at the same time, coaxial cables connecting television receivers to each other in a given area could eventually provide links between homes and businesses.

Banks are typical data users. Their online networks are used to update savings, demand-deposit, and installment loan accounts. Typically, the central computer interrogates the terminal at each branch office, requesting a response. Another requirement is transmission between the data center and an operations center some miles apart. Remote job entry sites would fit into this category. Many of the signals are carried in multiplexed form, where several lower rate channels are combined into a single high-data stream. The same type of service is often required by businesses for controlling finances, inventories, shipments, and so on.

Technical problems come in at this point. To state one, applications involving RJE line printers designed to run at 300 lines per minute will not operate at desired speeds if the communications channel's throughput is less than 4.8 kbps. Design perspectives thus become fundamental, and they must reflect use criteria. One of the more basic criteria is to make a clear distinction between the *volume* of traffic to be carried on the data communications links and the *response time* required by the operators or the various terminals.

Much too often, teleprocessing systems have been designed using the historical precedent of teletypewriter networks, which concentrated on the total *volume* of information to be transmitted over a relatively long period of time. The facts that a single message often takes several minutes to be transmitted and messages must be queued have been of secondary importance; however, with online teleprocessing systems, response-time considerations demand a whole new approach in engineering communications circuits.

Facilities with data-rate capabilities significantly higher than teletypewriter speeds are becoming requisites of these systems. These facilities are attainable by efficiently multiplexing wideband service readily provided by existing common carriers and now offered by CATV. These are the technical facts behind a financial institution's decision to enter a joint pilot venture with CATV technology. Let us take again Bankers' Trust to illustrate. The bank parallels the existing 50-kbps wideband lines between its BankAmericard division and the data center with a Manhattan CATV channel and offers "live" data and its existing multiplex channel equipment. CATV would develop the necessary interface devices and arrange to pull cable on both sides.

The introduction of point-to-sale (POS) terminals at merchant locations and the growth of branch bank automation within the last few years present an interesting application for the cable system. The fact that Manhattan Cable's data service is sold on a "no mileage" basis with minimal additional charges for multidrops makes it a viable alternative to the telephone network for POS applications.

The applications possibilities can better be brought into focus when we recall that in the 1950s, the majority of remote data processing networks comprised low-speed teletype circuits. Technological advances in equipment design, however, have necessitated increased data throughput requirements, often to megabit per second transmission. In the early 1970s, 2400 bps was the maximum throughput on private leased-line circuits, but by the mid-1970s it was possible to transmit 4.8 kbps on dial-up lines and 9.6 kbps on private-unconditioned leased lines, then 19.2 kbps, and more recently 56 kbps.

The data transmission environment in New York City is almost totally analog, interconnected by telephone circuits. With the enormous growth of data processing and online terminal usage during the 1980s, the telephone network is having difficulty keeping pace with the ever-increasing network demands.

Chapter 12

SATELLITE SYSTEMS: THE COMING BUSINESS

INTRODUCTION

During the 1980s we shall probably see major breakthroughs in waveguide communications and fiber optic techniques, as we have discussed in preceding chapters, but no new technology will have more of a far-reaching social, economic, and industrial impact than satellites. Satellite communications are no longer the technology of the future; they are here today, and their use continues to grow dramatically.

Artificial earth satellites were first successfully launched into orbit in 1957 (Sputnik), but their first public impact on television broadcasting was made with the launching of Telstar I into a low-altitude elliptical orbit in 1962, carrying an active transposer. Television pictures had previously been bounced from a passive satellite (Echo I), a metal-covered plastic sphere 30 meters in diameter, which soon became deformed from punctures in its aluminum–Mylar skin.

Telstar I was developed by Bell Telephone Laboratories, had a comparatively short operational life, and was used to exchange a large number of television programs between North America and Europe. The period of mutual visibility varied from orbit to orbit. To increase this period a British earth station was sited near the Atlantic coast at Cornell and had to be capable of tracking the satellite in its low orbit.

Yet Telstar I provided broadcasters with a completely new facility: live television relays across the Atlantic. (Prior to this, the only practical system, other than physically transporting tape or film, had been slow-scan systems for sending short news items via the transatlantic telephone cables.) A year later, the first relay satellites developed by RCA were launched into a higher orbit. They provided longer periods of mutual visibility, but such a facility was still way off compared with today's standards.

In the same year (1963) an alternative approach, the geostationary satellite, was actively promoted by Hughes Aircraft. The idea had originally been suggested in 1945 by the British engineer and well known science fiction writer Arthur C. Clarke. Clarke noted that for space satellites there was a unique orbit at 36,000 km above the equator where a satellite would appear to remain stationary by reason of its synchronism with the earth's rotation. He also postulated that a worldwide system of telecommunications could be designed using only three satellites carrying microwave repeaters.

During 1964, Hughes Aircraft, with the support of NASA and the Department of Defense, built and launched a series of synchronous satellites; Syncom I and II carried microwave transposers of restricted bandwidth; Syncom II was used during the 1964 Olympic Games to relay television pictures from Japan to the United States.

The first operational communications satellite "Early Bird," later called Intelsat I, was launched in 1965. It carried two transposers, which could be used for either television or multichannel telephony to carry commercial communications across the Atlantic. Originally, only black-and-white television was intended, but the system proved capable of relaying color transmissions.

Improvements in the art of satellite communications come in terms of bandwidth and switching. Many of the advanced telephone systems available through carriers (or independent manufacturers) lend themselves to the operation of satellite services. In this sense, the main telephone switch is used to control and operate in conjunction with one or more subsidiary nodes.

Satellites operations may be linked to the main switch through leased lines (foreign exchange or tie-line trunks) or customer-owned communications facilities. Incoming calls to a satellite facility will then be automatically routed to the nearest geographic location (satellite) for proper disposition.

Technically, the satellite will operate in the range 12 to 14 GHz and will communicate directly to rooftop antennas placed on the customer's premises. Current plans are to offer the user a minimum channel of 56 kHz; this compares favorably with today's leased-line channel that can carry a maximum of 9600 Hz (high-cost modems). Furthermore, the system:

- Will be dynamically flexible
- Carry integrated voice, data, text, and image traffic
- Offer wideband channels that can be multiplexed as the user wishes

The user will be able to rent one wideband channel costing roughly the same as a leased voice-grade line. This line, however, will be capable of carrying 16 data channels at 24 kbps or one wideband for voice, text, or data.

While technology takes big strides forward, how fast the satellite systems expand will depend not only on the facility itself but also on tariffs. To a very large measure, tariffs for the new satellite business services have not been set. Worldwide, the rates will be determined by the individual telcos (whether private companies or PTTs), according to a tariff structure which will be distance independent. A 1-Mbps channel will therefore cost the same whether it is used primarily to transmit information from New York to Philadelphia or from New York to London.

To protect their major market, local PTTs are unlikely to adopt satellite rates that would undermine their terrestrial telecommunications investment. British Telecom, for example, admits that its satellite channel rates would be higher than charges for the longest national land line or microwave link of equal bandwidth. On the other hand, it reports that satellite services would be cheaper than the longest international leased lines.

COMMERCIAL SATELLITE COMMUNICATIONS

The era of commercial exploitation of satellites opened with the Communications Satellite Act of 1962, which set up ComSat as the U.S. "carriers' carrier." This was followed by the 1964 formation of the International Telecommunications Satellite Organization under an interim operating agreement of the member states to own and operate a system of communications satellites for international telecommunications. Its growth rate in participating countries and earth stations is shown in Figure 12-1.

In 1972, the FCC authorized American common carriers to construct and operate satellite systems for domestic telecommunications. Initially, AT&T was prohibited from furnishing new private line services via satellite, but it retained its monopoly on long-haul message telephone service in the continental United States.

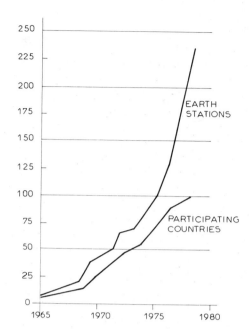

FIGURE 12-1 Growth rate of the Intelsat network in terms of participating countries and earth stations over the timespan of 15 years.

Western Union launched *Westar*, the first domestic system, in 1974. RCA and AT&T quickly followed with *Satcom* in 1975 and *Comstar* in 1976. These systems provide a wide range of domestic private line and message telephone services to the continental United States, Hawaii, Alaska, and Puerto Rico.

Table 12-1 outlines the operating satellite communications facilities, including those in advanced planning and R&D. As satellite science advances, telecommunications companies have to choose between either gambling on the latest invention or benefiting from economies of scale on earlier developments.

AT&T chose the latter course in 1980 by ordering a $138 million set of Telstar 3 satellites from Hughes Aircraft. Telstar 3 wrings more carrying capacity and longer life out of the tried and proven design of the 1970s. In 1983, AT&T hopes to launch the first of these Telstar 3s for its domestic telephone network to carry not only telephone conversations but also computer data, facsimile, and television.

Current technology calls for each satellite to contain one active and three spare wideband receivers, together with 10 communications channels. Each SBS (Satellite Business Systems) satellite, for example, will carry 10 transponder channels, with the power output of each 20 W and a usable bandwidth of 43 MHz. The data-carrying capacity of each transponder is 41 Mbps, for a total capacity of 410 Mbps per satellite. Another way to see the 410 Mbps is in terms of "X" computers.

TABLE 12-1   Satellite communications.

Operating or in Advanced Planning	Starting Date
1. LES Series (U.S.)	1965
2. USSR (Molniya II)	1965
3. DSCS (U.S.)	1966
4. Skynet (U.K.)	1969
5. NATO	1970
6. Telesat Canada	1973
7. Western Union (Westar)	1974
8. American Satellite Corp.	1974
9. ARS-6 (U.S.)	1974
10. Symphonie (Germany-France)	1975
11. CTS (U.S.–Canada)	1975
12. Comsat General (Marisat)	1975
13. Intelsat IV-A	1975
14. Norway/North Sea	1975
15. Algeria	1975
16. Malaysia	1975
17. Indonesia	1976
18. Philippines	1976
19. Sirio (Italy)	1977
20. Japan Telecommunications	1977
21. ESA (Marots)	1977
22. ESA (OTS/ECS)	1977
23. FltSatCom (U.S.)	1977
24. Japan-Broadcast (BSE)	1978
25. Brazil	1978
26. ESA/Comsat General (Aerosat)	1979

R&D Status (early to mid-1980s)

1. SBS	8. Andean nations (regional systems using private Intelsat transponders)
2. Arab System	
3. Australia	
4. European Broadcasting Union	9. Argentina
5. United Kingdom	10. Denmark
6. IMCO (Inmarsat)	11. Iran
7. India	12. Federal Republic of Germany

Say that a computer at 0.2 μs (a 200-ns machine) has 2 Mbps. Then, 410 Mbps means 205 computers working in parallel. This is internal traffic (for instance, memory dump); external traffic today never reaches this level.

AT&T's version will carry up to 21,600 simultaneous telephone calls instead of the previous limit of 18,000, and its expected lifetime will be 10 years instead of seven. It is a performance to be compared with Early Bird (Figure 12-2). Early Bird could handle only 240 tele-

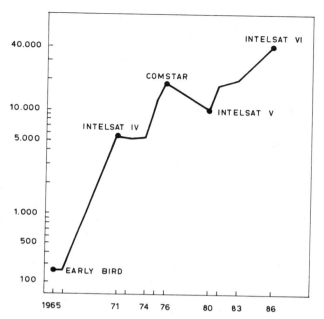

FIGURE 12-2   A 20-year timespan: growth in number of telephone circuits supported by a satellite.

phone calls and was designed to last just three years, although it actually lasted much longer.

AT&T Long Lines move to replace Comstar by 1983, to carry 4- and 6-GHz, 12- and 14-GHz, and possibly 20- and 30-GHz transponders. This system will be able to support high-capacity trunks between major cities.

By 1986, Intelsat VI, built for the International Telecommunications Satellite Organization, will become the world's most sophisticated commercial communications satellite. This drum-shaped, spin-stabilized unit will have twice the capacity of Intelsat V. It will be able to carry 33,000 telephone calls and four TV channels simultaneously. It will weigh about 4 tons at launch and measure 4 meters in diameter.

The first satellite for the nationwide communications system being established by Satellite Business Systems was successfully launched in November 1980 and is being tested in its final orbital position. (SBS is the partnership formed by subsidiaries of The Aetna Casualty and Surety Company, Comsat General Corporation, and IBM.) This company will provide a variety of high-speed digital communications services to meet the needs of organizations of varying size. Commercial operations starting in 1981 will serve a number of leading U.S. companies.

SBS—operators of the first fully integrated, wideband, switched satellite network—launches its domestic Telstar-type satellites in the 12- and 14-GHz band. Its business objective is to install private communications networks for:

- The Fortune 500 leading corporations
- Large public service institutions

throughout the continental United States. The market is estimated at 200,000 "equivalent two-way voice circuits" by 1985, which means some 10 satellites of the Telstar 3 variety.

The satellites' payload control capacity is shown in Figure 12-3. In terms of performance, satellite data communications have exhibited excellent service performance, high data-rate capability, low bit error rate (10^{-8} or better), and very good reliability (typically 99.99%). But if satellites are to become the world's long-distance telecommunications means, they must keep costs down.

Submarine cables are trying to regain their former dominance. The new undersea cable, which will operate on the transatlantic route in 1983, can handle 4000 telephone calls. Figure 12-4 compares number

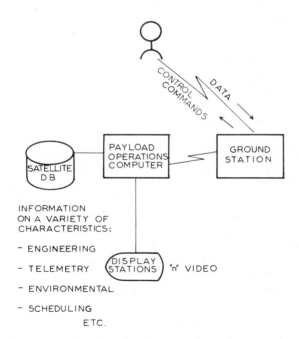

FIGURE 12-3   A satellite's payload control facilities supported by the ground station.

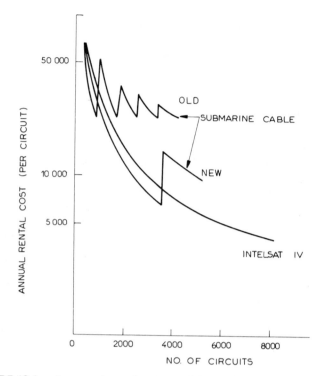

FIGURE 12-4   A comparison of the number of circuits to annual costs for a satellite system (Intelsat IV) and the old and new submarine cable solutions.

of circuits to annual costs. Optical fibers will soon extend the capacity of cables beyond that level.

The launching vehicle is a key issue. Telstar 3 was to have been carried into orbit by the Space Shuttle, and the plan will probably materialize because the Space Shuttle has been successfully launched and Telstar 3's 1983 deadline is still some time away. Possible alternatives were the Atlas Centaur, Delta, and Ariane rockets.

The launching vehicle is so important because the technical trend has been to orbit larger, more complex antennas which can send and receive transmissions more efficiently, thus extending the life of a satellite by reducing power consumption.

The corresponding reduction of the area the satellite could cover with its transmission is a problem. It has been addressed by bouncing transmissions from earth to one satellite, back down to earth, up to another satellite, and then down to the final earth destination. This process will eventually be solved through a switching solution in space: either a platform or a system of dedicated satellites.

SATELLITE SERVICES

There are basically two types of satellite services available. The first transports data over major trunk facilities:

- City to city
- Domestic or international

through normal, terrestrial interconnective means to the regional earth stations (Figure 12-5).

With current technology, for the majority of companies this is a valid solution to the high speed, interoffice communications. It was given a boost in December 1980, when the FCC gave nine firms permission to build and launch more than 24 communications satellites until 1985. Several of these firms offer diverse communications services. In fact, only two of the companies, Communications Satellite and Hughes Communications, are exclusively satellite networks.

Nineteen months later, in July 1982, the Federal Communications Commission authorized the first sales of channels on communications satellites, a move that will allow private businesses to buy, rather than lease, slots on the satellites. It did so by approving plans by Hughes Communications, RCA Americom Communications, and Western Union Telegraph to sell a total of more than 100 satellite channels, known as transponders.

The second approach involves earth stations located on customer premises and able to provide DTE-DTE (data terminating equipment) communications. The important advantage of this end-to-end system is that it permits the wide dissemination of high-speed data transmission to multiple locations in a network without needing terrestrial distribution facilities. This serves:

1. The need for high-speed text and data transmission to regional offices—printing and distribution plants for newspapers and other mass media being one example

2. The requirement for high-rate document distribution (facsimile, et al.) to meet electronic mail requirements

3. The transmission of large volumes of data among various computer systems on a broadcasting or narrowcasting basis

4. Special types of requirements, such as direct access communications to offshore oil rigs, ship terminals, and other remote geographical locations

As an example, Figure 12-6 presents a current solution adopted by the Dow Jones wideband facsimile network. Packet-broadcasting systems

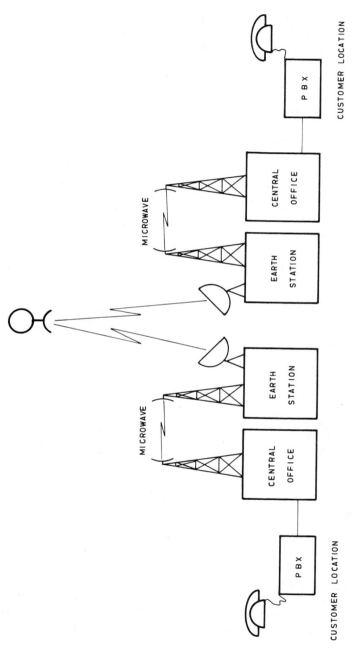

FIGURE 12-5 A satellite system from PBX to PBX through central offices, microwave links, earth stations, antennas, and the satellite itself.

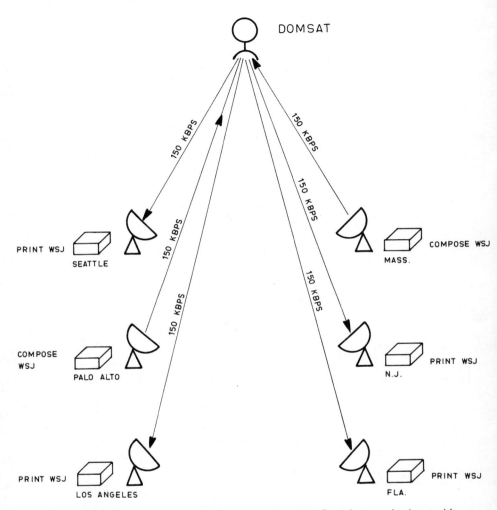

FIGURE 12-6 The Domsat solution adopted by Dow Jones and using a wide-band facsimile network.

are an attractive possibility given the efficient use of space-segment satellite resources.

Table 12-2 presents another example: a 1982 list of SBS users with dedicated earth stations. Notice that these companies include insurance, banking, merchandising, electrical/mechanical engineering, and so on. Voice, data, and teleconferencing (in that order) characterize the applications.

This can be said about capacity projections: The Intelsat global

TABLE 12-2   SBS user companies with dedicated earth stations.

	Voice	Data	Telecon-ferencing	Current Earth Stations	Earth Stations Projected for 1984/85
1. Isacomm				10	50
2. Boeing	X	X		3	5
3. Crocker National Bank	X	X		3	23
4. Dow Chemical	X	X	X	3	3
5. General Electric	X			3	3
6. General Motors	X			3	10
7. Hercules	X	X	X	3	15
8. IBM	X	X	X	3	20
9. Allstate Insurance	X	X	X	2	6
10. J.C. Penney	X	X	X	2	12
11. Travellers	X	X		2	3
12. Wells Fargo	X	X	X	2	2

satellite system today provides the equivalent of over 20,000 two-way voice-grade circuits among 90 nations for long-haul overseas voice and data traffic, and television distribution. Such units of measurement can be swamped by the requirements of just a dozen companies in Table 12-2.

Even these projections should be looked at with great care, as technological breakthroughs like the Space Shuttle can change the quantitative basis very radically. A lesser but still significant development such as the Ariane rocket orbiting the European satellite *Heavysat* (weighing 900 kg) might become the forerunner of heavy-duty satellites, able to handle direct and semidirect transmission of TV and radio programs. (The accords for Heavysat have been taken between ESA, the European Space Agency, and EBU, the European Broadcasting Union.)

The basic criteria relative to service, economics, and overall performance must be established beforehand, and the technology available must be explored to this full capability. The design of satellites for commercial exploitation presently centers around engines powered by solar energy (supplying 300 to 800 watts) and launched by NASA's workhorse rocket launchers.

Table 12-3 identifies the generations of communications satellites from 1974. With the Space Shuttle capabilities for orbiting heavier satellites, the next generation will probably weigh 2000 kg (and be 42 X 54 X 15 feet), and provide 1800 watts of solar power.

In the current and projected design specifications, the communications payload includes transponders in three frequency bands (2 GHz, 4 and 6 GHz, and 12 and 14 GHz) to support the NASA and Advanced

TABLE 12-3   Three generations of communications satellites.

	GHz	kg	Watts
"0" generation	4–6	500–2000	300–800
(voice orientation)	12–14		
1974–1975			
"1" generation	4–6	2000	1800
(data communications)	12–14		
1980–1981	20–30		
"2" generation	4–6	4000–5000	4000
(TV broadcast,	12–14		
teleconferencing,	20–30		
electronic mail,	30–40		
voice, and data)			
1985–1986			

Westar services. In the projected mode of operation, each satellite can provide:

1. Twelve 36-MHz channels at 4 and 6 GHz

2. Four 250-Mbps time-division multiple access (TDMA) channels in the 12- and 14-GHz band, with the lower frequency for the uplink and the higher for the downlink

The new commercial satellite systems will employ a large number of relatively small earth terminals. Individual private line subscribers will have their own dedicated satellite communications link for high-speed data, text, and voice traffic.

A global design involves three broad regions of the world, identifies border areas, and assigns regional distribution to networks served by coaxial cable and optical fiber installations. This capitalizes on one of the most important advantages presented by satellite transmission: the visibility the satellite has over a large terrestrial area.

Such an approach permits multiple access by earth stations to and from the satellites. Yet, though this area is large, it still covers only part of the globe and there is at the present time no satellite-to-satellite direct transmission. Transmission and retransmission must be assured through earth stations. For that purpose, as Figure 12-7 demonstrates, the world is divided into three areas:

- Area 1: Pacific, including North–South America and Australia
- Area 2: Atlantic, basically Europe and Asia
- Area 3: Asia, with a central point the Indian subcontinent

This three-area organization not only answers globe-wide retransmission requirements, but it also reflects the historical evolution of satellites

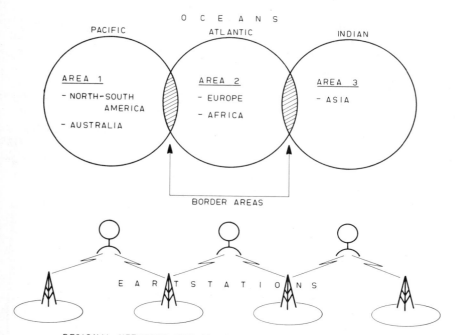

FIGURE 12-7   Pacific, Atlantic, and Asia—a three-area organization for global retransmission requirements through satellites.

as world communications media. Indeed, the first use of satellites focused on intercontinental distances. This approach has had a significant evolution during the last decade, as geostationary satellites are becoming a very important means for the constitution of national high-capacity communications networks.

Technology makes it feasible to steadily reduce the size of the satellite station; this leads to cost-effective networks of earth stations. Money center banks and large industrial companies are particularly interested in the use of earth stations at their major operating units with local networks bringing the voice, text, data, and image to the end user.

DOMESTIC SATELLITES

The use of domestic satellites for distributing television and radio programs is becoming a normal application in countries with developed terrestrial microwave systems. In the United States, distribution circuits and temporary links are now available via Western Union's Westar satellites and RCA's Satcom satellites. The Federal Communications Commission has authorized U.S. use of compact receive-only terminals for the downlinks from satellites which provide, for example, 15-kHz sound circuits compared with the 5-kHz circuits generally used in the American long-distance terrestrial network.

In this sense, a domestic satellite system becomes the logical extension of a country's basic communications understructure. This is the case in Indonesia, which has a geostationary satellite located at 83°E for relaying television signals to about 40 stations throughout the 26 Indonesian provinces. The earth satellite, instead of functioning merely as a relay to either single or multiple earth stations, itself forms the television or radio broadcasting transmitting station.

As Robert C. Hall, SBS president and CEO, was to suggest: Satellite communications has opened up the dawn of the fully distributed wideband era." This provides the opportunity to escape the limitations of voice-oriented data links, and means an explosion of applications for advanced communications. (SBS is by no means the only player: AT&T, American Satellite, RCA, Americom, Western Union, and Xerox are all planning new or expanded satellite services; GTE, Southern Pacific, and Hughes Communications have recent FCC authorization to proceed toward implementation of their own domestic satellite systems.)

The design of increasing satellite capabilities is an outstanding technological development. Theoretically, a single synchronous satellite would be capable of beaming line-of-sight signals directly to approximately one-third of the surface of the earth. However, for practical purposes, limitation of coverage is desirable, as it permits the use of more realistic transmitter powers.

Earlier it was thought that the limitations of solar cells would preclude direct broadcasting from satellites; consequently, the possibilities of developing compact nuclear generators or fuel cells were considered. The exploitation of nuclear power in space was subsequently felt less likely because of the possible risk to the public and to on-board electronic components.

Frequency allocation is a technical and political problem. International agreements reached at the ITU World Administrative Radio Conference for Space Telecommunications (held in Geneva in June–July 1971) restricted frequency allocations for space broadcasting to:

- Parts of Band V (620 to 790 MHz), which is already fully utilized for terrestrial broadcasting
- 2500 to 2690 MHz (not all available in Region I)
- 11.7 to 12.5 GHz
- 22.5 to 23 GHz (Region 3—Asia and Oceania only)

- 41 to 43 GHz

- 84 to 86 GHz

Subsequently, a 1977 World Administrative Radio Conference produced a "World Agreement," and a plan for Regions 1 and 3, for the 11.7- to 12.5-GHz band. Let's recall that radio broadcasting began around 1 MHz, but soon involved *Empire* services between 6 and 16 MHz. Early high-definition television called for 40-MHz reception, but in 1955 the British ITV with Band III put television in the region of 200 MHz, while the 625-line color services have raised the limit to 470 to 859 MHz. Yet never before has a single increase reached such a threshold as would the introduction of 12-GHz satellite television.

The subject of satellite television is of great actuality as it beams transborder. This issue is far more important to Europe, where one single satellite can easily beam to four different nations, and where the TV stations are government controlled while television satellites are being projected.

The first known effort of a satellite for experimental television broadcasting purposes was a Japanese engine known as "BSE." This was successfully launched during 1978 on a U.S. launch vehicle: it had two sets of 14/12-GHz 100-watt transponders, placed over the equator at longitude 110°E; the television uplinks were on 14 GHz, and the downlinks were on approximately 12 GHz. The satellite (with a three-year design life) provided a power flux over the Japanese mainland; its solar array assumed about 1 kW of power.

Such experiments have demonstrated that the downlink (satellite to earth) is significantly more demanding in its requirements for good receiving installations than is the uplink (earth to satellite) where cost is of less importance. The limiting factors in satellite reception are the aerial gain and thermal noise (noise temperature) of the receiver. The aerial receives from the sky unwanted noise energy which increases rapidly at low angles of elevation. Ideally, the first stages of a receiver should have not only a low noise temperature but also a sufficient gain to make the noise succeeding stages insignificant. These technical subjects will be more easily solved as we increase the size and capacity of the satellites practically feasible after the Space Shuttle's successful launching.

Between 1965 and 1980 Intelsat grew from 240 telephone circuits to its present capacity, doubling every two years. ComSat predicts over 400,000 equivalent two-way voice circuits in service by the year 2000, but a projection made by the Bell System for its own Long Lines traffic indicates a total of some 2,000,000 "satellite eligible" circuits by 2000 —up by a factor of 5.

Let's say that a "satellite eligible" circuit is arbitrarily defined as one over 1000 miles long. Other educated guesses as to the total expected demand for satellite circuits *for present uses* in 2000 (including domestic, regional, and international traffic) bring the needs up to 3.5 and 4 million circuits.

The latter hypothesis calls for over 10 times today's capacity; with new uses, the total demand in 2000 may be 15 or 20 times today's capacity. Not all will be in satellites—fiber optics and millimeter waveguides will absorb some of the demand—but a good deal of the total communications capability will probably be in orbit.

The able use of broadcasting services calls for forward-looking, efficient study able to cover a broader area of activities: from satellite design and earth station capabilities, to system maintenance solutions and international agreements.

Chapter 13

BEYOND THE LAUNCH VEHICLES

INTRODUCTION

There are trade-offs that can be made in designing a satellite system. Some of the functions handled through an earth station can nicely be designed into the satellite, thus making the installation of hundreds of thousands of earth stations simpler and less costly. This, however, presupposes heavier satellites, a practice reasonably inhibited by the fact of launching them from earth. It also calls for more advanced spacecraft technologies, such as stabilization and power, which in turn determine total spacecraft weight and communications capacity per unit weight.

Spacecraft weight can have an appreciable influence on launch costs, thus affecting overall system economy. The solution lies in launch vehicles such as the Space Shuttle. With the Shuttle, weight may be less of a problem in spacecraft design.

Further possibilities may develop if the Shuttle proves cost effective compared to today's expendable launch vehicles and if technological advances make it feasible to recover satellites, perform maintenance, relaunch, and reuse them.

Throughout the range of these developments, systems studies will be at a premium and technical characteristics need detailed study. For instance, the downlink from satellites to earth imposes requirements due to the limited flux permitted from the satellites. Given the capability of building sophisticated gear on earth, the uplink poses lesser challenges, and it is possible to inject television transmission into a satellite network from compact, transportable, mobile earth stations.

In 1972, for example, a compact earth station was air-lifted in two Boeing 727's to Peking Airport and used to relay pictures of President Nixon's visit to China. These pictures reached the United States via the Intelsat Pacific satellite working at the Jamesburg earth station near San Francisco. For the 1976 Olympic Games, Teleglobe Canada (the Canadian overseas telecommunications organization) installed a temporary earth station on Mount Royal, near Montreal, to augment their capability to distribute the coverage of the Games worldwide.

There is a distinction to be made between the demountable stations, those which are more truly portable, and the stable network of earth stations needed to implement a satellite-supported worldwide business system. The truly portable and fixed networks can both be components of the latter, needed, respectively, for handling small module contribution facilities and aimed at dedicated distribution circuits or direct broadcasting to small rooftop receiving terminals. Both classes raise problems of control and administration which the operating agencies must resolve.

THE SPACE SHUTTLE

The Space Shuttle, or Space Transportation System (STS), represents the first major change in the concept of launch vehicles since the space program began.

Until now, the focus has always been on launching an object into space, where it would perform some feat. Although these objects were of increasing size and complexity, starting with the Sputnik and culminating with three astronauts and their moon lander in 1969–1970, the basic launching configuration was:

- A one-time throwaway rocket
- A single payload

But the emphasis now shifts from the payload to the launcher itself. The primary interest is on the economics of space: the greatest number of launches for the fewest dollars spent. This requires versatility in launch capability to satisfy the great variety of potential users. It also calls for cost effectiveness, as new markets can be exploited only if the cost is right.

In this sense, the Space Shuttle signals the beginning of an era of commercial use of space where multiple payloads are regularly lifted into orbit thanks to an inexpensive and versatile launch capability. As such, it will undoubtedly expand space technology applications and make it possible for small as well as big enterprises to launch business ventures in space.

Once the STS is operational on a regular service basis, it is likely that previously unforeseen uses will emerge among industries now seemingly far removed from space. In a symposium held in the United States, it was projected that by 2000, business conducted in space could generate revenues totaling $30 billion.

For as little as $3000, small business foundations, universities, or individuals can launch their project or experiment abroad the STS while a dedicated mission (a mission with only one user) will cost about $21 million. And this is cheap!

Relating STS costs to the expandables, the original "Thor" rocket costs about $10,000 per payload pound; the Shuttle cost will likely be about $1500 per payload pound, and, as bigger spaceships are built, this cost will drop.

Looking at the possible space industries, we can distinguish: first, public communications to fit the growing need of private communications and data transmission, and second, solar power as the Space Shuttle will make it possible to construct huge solar power stations in space. (Initially, such stations will be financed by governments because of the tremendous costs, but it is inevitable that private industry will become involved at some time.)

Special manufacturing is a third business opportunity which is attracting considerable attention. It is possible, for instance, to make perfect ball bearings in the weightlessness of space. Other opportunities include processing certain metal alloys and chemical compounds which are difficult or impossible to process on earth. Space factories will some day even manufacture finished products that will be used in constructing other space projects, such as solar power stations.

Meteorology constitutes a fourth business opportunity. As more advanced weather satellites are developed and launched, it will be possible to make more accurate long-term predictions. Meteorology has long used weather forecasting approaches; with satellites it gets its needed vital tools.

Health care is a fifth business opportunity, as space hospitals will make possible advances in medicine such as spinal cord fusions, another feat impossible on earth. Severe burn victims can be treated much better in the weightlessness of space than they can be here on earth.

A sixth area of business opportunity results from the projected advances in geology. Geologists will become more effective in their never-ending search for minerals and energy resources.

A seventh business reference pertains to leisure, travel, and colonization. Within the foreseeable future, private citizens will travel to vacation at space hotels; this could happen by the end of this century and will both be an aftermath and a promotion of full-time space living. It is certain that huge space stations will be built to house the space labor force that will work on many of the described projects.

We have come a long way since 1957. Imagination, coupled with confidence in this rapidly expanding, highly technical industry, has produced great advances and even bigger challenges.

All these are good reasons why it matters so much that the 1981 STS flight was successful. Twenty years, to the day, after Russia's Yuri Gagarin made the first manned orbital flight, America's John Young and Robert Crippen became the first people to leave this planet in a spacecraft that can be used again and again. Their mission has been, in practical terms, an even more important milestone than Neil Armstrong's "giant leap for mankind." The moon landing was symbolic; the real significance of the space age is not so much that it will extend human frontiers, but that it will make life more pleasant in our global village.

The space age, indeed, really began with the Space Shuttle; yet its earlier approximations have provided many benefits to humankind. It has made intercontinental telephoning commonplace, permitted different nations to watch the news as it is being made, and helped environmentalists, fishermen, and farmers, as well as oil and mineral prospectors. By giving more advanced and more exact warning of hurricanes and tornadoes, the space age can save lives. The Shuttle promises to extend these benefits and to add new ones.

When the STS is fully operational, perhaps in 1985, it will make about 40 flights yearly; be able to carry into space new, improved, large satellites at about one-fifth the current cost of launching much lighter structures; and allow heavy satellites to transmit directly to ground stations on any rooftop. The real international telecommunications age will then begin.

SATELLITE ANTENNAS

Present-day satellites use transponders to produce the bounce or reflection effect. A transponder is a combination receiver/transmitter that acts like a repeater in that it accepts and returns a signal transmitted from the earth.

It is common to use different transmission frequencies for the up and down directions to prevent interference between the two directions. This is a concern because the reflected signal will reach not only the intended receiver but also the original transmitting antenna as well as all other antennas within an area of several million square miles: it is easier to hit a stationary target than a moving one. Technologically, a solution is reached by carefully choosing the satellite orbit speed, which is related to orbit altitude, so that to an observer on earth the satellite will appear stationary: the angular orbit speed matches the earth's rotation.

This is the geostationary or synchronous orbit to which reference was made in the preceding chapter. Because of the earth's rotation, such an orbit can be achieved only at a point over the equator. For obvious reasons, a satellite with a geostationary orbit cannot be used to transmit information to the other side of the earth. With an orbit altitude of 23,000 miles (36,000 km) around an earth with a 4000-mile (6400-km) radius, the theoretical maximum coverage possible is approximately one-third of the earth's surface (as we have already noted in the reference to the pioneering work of Clarke.)

Still, as a practical matter, the coverage or "footprint" of a satellite is substantially less. When information is to be transmitted beyond the footprint, a satellite "hop" is used. The signal bounced from a satellite is again bounced from a ground station to a second satellite, which sends the signal to a receiving station within the second satellite's footprint. The ground station used between the two satellites must, of course, lie within an overlapped area of the two footprints.

Satellite antennas can be *omni* or *directional*. An omni antenna radiates signals in all directions nearly equally. This can waste power in the sense that power is sent to areas where there are no ground stations. A directional antenna concentrates its radiated power toward the geographical areas where it can be used, but because of the usage of the directional antennas, a satellite's footprint is generally not the perfect circle on earth that one would expect.

Furthermore, there is an inverse relationship between transmission frequency and antenna size. Commercial FM broadcasts are of a much higher frequency than AM broadcasts. FM radios usually are sold with the recommendation that antennas be kept at 30 inches: if the length is reduced below 30 inches, reception would continually deteriorate down to zero. Similar discussion can be applied to dish antennas to determine minimum optimum size. Dish antennas are also designed to gain: they will actually enhance the signal power rather than simply grab as much as possible.

Other design problems must be overcome. The receiving antenna must be pointed with an accuracy of better than 0.5° toward the satellite. When this figure is related to an average UHF aerial, it becomes evident that installations will require much care. Means of adjustment must be provided, and it is possible to line up a satellite antenna by observing the reception rather than by dead reckoning. Further still, no matter how effectively the receiving antennas have been designed, the long-term performance will depend on the care with which they are installed and maintained. The view of the satellite must be unobstructed; screening, for example, may be a serious problem in an area with large buildings or tall trees.

The next generation of earth stations and satellites will necessarily incorporate features of the preceding one. An example is that of multiple frequency bands to serve the developing earth terminals network (Figure 13-1). This means:

- 4/8 GHz and 10/14 GHz into existing earth terminals to satisfy increasing demand

- In addition to the lower bands, 20/30 GHz to meet projected needs

Apart from investment preservations, there is economic pressure to combine services to multiple sets of users on one satellite. The next critical design factor is the digital communications discipline.

The time-division multiple access/demand assignment (TDMA/DA) is a technically valid transmission technique chosen for multiplex digital

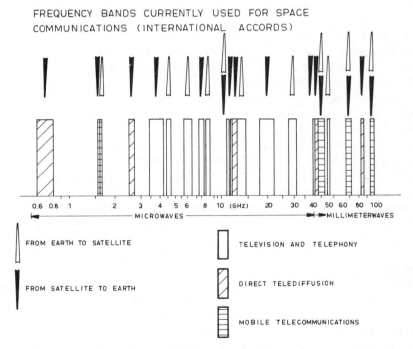

FIGURE 13-1 The multiple frequency bands serving the developing earth terminals network (earth to satellite and satellite to earth).

bit streams sent by the earth stations. TDMA/DA, for instance, is a most efficient method for optimizing use of the 25-Mbps bandwidth of the Telecom-1 transponders, but it requires extremely accurate synchronization between individual customer earth stations and the network. In general:

- TDMA (Figure 13-2) is a satisfactory solution if all satellite downlink signals must be received by all the earth stations, while
- SSTDMA (spacecraft-switched time-division multiple access) is a newer technique that overcomes the problems of broadcast waste.

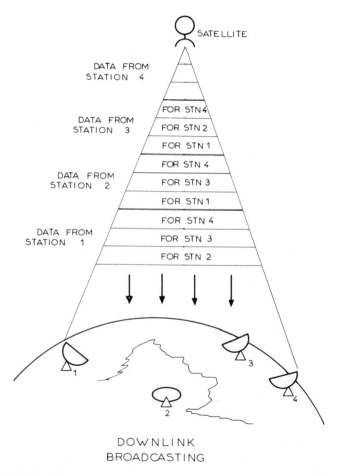

DOWNLINK
BROADCASTING

FIGURE 13-2 Uplink and downlink, earthstation to/from satellite using TDM technology. The downlink is broadcasting.

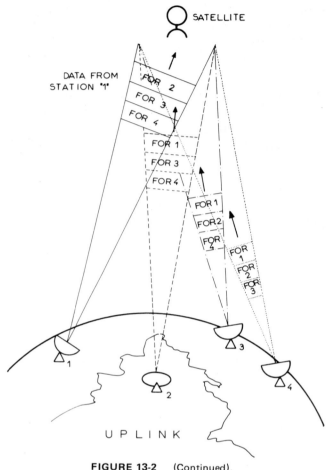

FIGURE 13-2 (Continued).

New advanced satellite antennas can aim all their power at a single station. Separate beams will be directed from the satellite to the various receiving stations. In a point-to-point SSTDMA system, all stations can transmit and receive data practically all the time (Figure 13-3).

Station operation in the area-coverage SSTDMA system is intermittent, since all stations in the area must share a single communications uplink and downlink. A satellite switch will rapidly change the connections between antenna beams so that:

- Each beam is connected in sequence to each other antenna beam.
- Each station can efficiently communicate with each other station via satellite.

In this way, area-coverage SSTDMA operation is similar to TDMA operation where all stations in the system share a single communications channel.

Advances in digital memories, phase shifters, and solid-state radio-frequency (RF) amplifiers make it possible to form an independent beam for each signal burst, directing the energy at the receiving station. This is satellite technology; it must be matched by earth station and maintenance developments for efficient and reliable system.

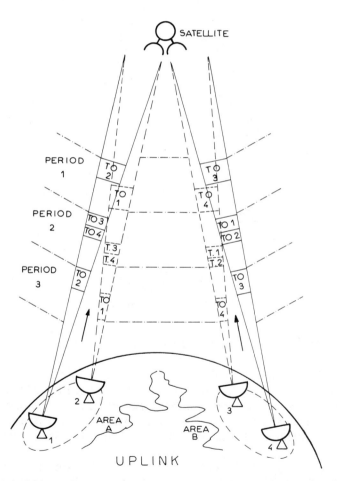

FIGURE 13-3   Uplink and downlink using spacecraft switched area coverage (SSTDMA). The bursts from the transmitting stations are interleaved.

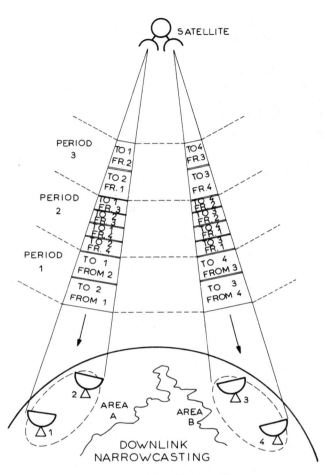

FIGURE 13-3   (Continued).

SATELLITE USAGE AND DAILY SERVICE

As the impact of satellite-supported communications services grows, we will experience a better facility, of lower cost per unit of transmission, with many more broadband channels. But we are also going to encounter more technical problems if we keep working with protocols established for a different form of supports. The following paragraphs say "why."

As of January 1982, AT&T began freely intermixing satellite and terrestrial communications paths in its Long Lines operations. This brings under perspective the propagation delay which can affect bisync and other ARQ (automatic request repeat) transmissions. This new situation has an impact on both users of the dial-up network and users of leased lines.

As a consequence of this change in policy, users employing bisynchronous and other protocols will need to undertake a software upgrade or else experience a significant percentage of degradation in efficiency of datacomm throughput, though at the present time Long Lines uses about 23,000 interstate satellite circuits, out of a total of 485,000 interstate AT&T circuits of all kinds.

It takes approximately half a second for data to be beamed 22,000 miles to a communications satellite and then relayed back to earth. More precisely: 0.24 s from earth to satellite and the same time for satellite to earth. This delay results in two types of problems for datacomm users.

One problem is the echo suppressors employed in telephone connections more than 1500 miles long. They would interfere with the transmission of data over the circuit, as they include disablers which are activated by a special signal from the transmitting modem. Yet, echo suppressors are necessary on terrestrial circuits as they permit the same lines to be employed for both voice and data communications.

On satellite circuits, the half-second propagation delay means that the turnaround time required to disable the suppressor becomes quite large. This problem may be solved by introducing hardware changes, but a number of technical and financial questions still remain to be solved.

A second and bigger problem concerns the transmission acknowledgments necessitated by the bisynchronous or other protocols frequently required by remote job entry terminals. Such terminals must receive a signal acknowledging that one block of data has been received before sending the next block. A half-second holdup every time this procedure occurs would chop the net throughput of terminals operating at 4.8 kbps down to 400 bps.

A solution is to provide a unit that emulates computer acknowledgment for received data. American Satellite, for instance, offers a Satellite Delay Compensation Unit (SDCU) which is provided as part of its data service offering. ASC suggests that the SDCU on satellite links can offer the user a 95% throughput.

Another option is a rather extensive software change with packet-switching protocols like High Level Data Link Control (HDLC) and Synchronous Data Link Control (SDLC). These require fewer acknowledgments and transmit larger blocks of data. Hence, they can reduce problems resulting from the usage of BSC.

So much for the bad news. Now the good news in terms of service to the user. The following example may be typical of datacomm implementation in this decade.

Faced with the need to upgrade data communications among five geographically separated data centers, Metropolitan Life Insurance Company initiated a comprehensive satellite network, using American Satellite and AT&T equipment, able to replace terrestrial links. Starting in the late 1970s, Metropolitan engaged American Satellite to provide a number of satellite links between four remote computer sites and its home office in New York.

As it currently stands, this high-speed satellite network comprises five nodes: computer centers in the home office, New York; Scranton, PA; Greenville, SC; Wichita, KS; and a regional office in San Francisco. Nodal interconnection is accomplished through 56 kbps American Satellite channels.

There are two AT&T terrestrial 230 kbps links between Scranton and New York. Earth stations are installed on Metropolitan premises at Scranton, Greenville, and Wichita; the San Francisco and New York nodes being supported by American Satellite-shared commercial earth stations.

In this network, IBM mainframes communicate with other IBM mainframes and with Honeywell Level 6 minicomputers. Where BSC (bisync) protocol is used, Metropolitan employs the ASC satellite-delay compensation unit to avoid the acknowledgment slow-down due to the half-second transmission delay in satellite communications. Where Metropolitan has Honeywell L6s talking to other L6s over satellite links, it uses the HDLC protocol.

With packet switching, a large block of data is transmitted before an acknowledgment is sent. The larger the block size, the fewer the acknowledgments needed to be transmitted back to the host—and therefore the lesser the impact of the transmission delay inherent in satellite-based communications.

TECHNICAL AND POLITICAL PROBLEMS

The power, universality, accessibility, and transmission range of satellites may make them the highways of the future. Their wide bandwidth will see to it that they become ideal carriers for text, data, voice, image (including television), electronic mail, and teleconferences.

Because the links between computers and communications are so great, mass access to computers and communications will be like access to electricity, but with the difference that the current is not inert: it contains information, hence power.

These are fundamental, background reasons why work in process can radically alter tomorrow's sociological, financial, industrial, and educational structures. However, there are challenges to be faced; one of the issues that requires international agreement is the spectrum/orbit slot problem.

As no two users can operate simultaneously in any part of the electromagnetic spectrum without interfering with one another, its segments must be allocated in a way to prevent interference. International bodies such as the ITU (International Telecommunications Union) work on that purpose, but conflicts do exist.

Although there is physically ample space in the geosynchronous orbit, there is also a limitation on how closely satellites can be spaced. Because of the need to space satellites, orbit slots and frequencies must be assigned (Table 13-1). At 12/14 GHz, the required orbital spacing is 4° of arc. There is plenty of room in the 30- to 40-GHz bands, with only 1° of separation, but there is also a major problem which affects communications systems at high frequencies: rain.

Getting through a thunderstorm may take as much as 4 to 70 times the power; therefore, the link power calculations must include a rain margin, enough extra power to face rain at the earth station. The amount of rain margin depends on the reliability of service required and other factors:

- Data transmission and telephony call for very high reliability and no outages.

- Television and facsimile might tolerate a delay until a bad rain storm passes.

Repeated transmission is one of the technical solutions. Transmitting a digital burst twice has the same effect as using twice the power. Yet technical solutions will take time, and changes in technology or in approach cannot be abrupt, even in a brand-new science such as satellite communications.

The third generation of satellites (large and powerful) will probably be put into geostationary orbit by the Space Shuttle starting in the mid-1980s. They will require less powerful earth stations and very simple receiving dishes.

TABLE 13-1 Orbit slot availability.

Service	Band	Nominal Spacing (deg)
Telecommunications only	4/6 GHz	4–5
Telecommunications and	12/14 GHz	4–5
TV broadcast		8–10
Telecommunications only	30/40 GHz	1
TV Broadcast only	30/40 GHz	1

1. Depending on their power, orbit, and angle, satellites can cover a whole continent or both sides of an ocean.

2. Send/receive earth station costs can be reduced to $200,000 when a mass market develops.

3. Experimental work indicates that receive-only satellite dishes can be produced for a unit cost of only $70.

This leads to the thought that if IBM or any other corporation dominates satellite transmission, the company could transcend the role of a mere manufacturer. It can become one of the great world regulatory agencies; consequently, it is vitally important to develop and implement valid international standards, planning and controlling world communications to benefit all. This is also urgent because the new generation of communications satellites will be launched between 1983–1986, and this will be followed by another generation of satellites which will capitalize on the lessons of the first two.

By all probability in the middle to late 1980s there will be available high-power communications satellites, competing with earth-bound networks. The pace of competition will be tremendous. Slowly the salient problem will move from a question of satellites, ground services, and the privately or publicly offered databases to acute competition between European and American computers and communications centers, with the latter having the edge.

Satellites will aid poor countries by offering many a means of introducing a telephone service cheaper than burying cables in the ground. In industrial countries the cost advantage of satellite calls will apply shorter and shorter distances.

All this has to be seen in connection with computer power, which will spread to every home and every desk through microprocessors, while automatic retrieval on microfiles will get increasingly generalized through optical disks.

Computers will play a surprisingly minor role in all these events; it is the combination of computers and communications that will have the major impact. Tariffs and regulations will affect tremendously the way the computers and communications power struggle will develop, and it has already been emphasized that the old approaches to regulation could be destructive.

If broadcasting is forbidden in Europe outside the PTT authorities, European offices of multinational companies could use receive-only rooftop dishes to get large volumes of data from U.S.-held databases. A new concentration of power (through information) will develop.

With bulk data transmission from the United States to Europe (via the SBS and other satellites) much cheaper than the equivalent broadcast from Europe to the United States (via the PTT and IRC), most corporate databases of both American- and European-owned multinationals might be shifted to the United States with more traditional radio links used to transmit the short ack (acknowledgment) messages.

Commerce, banking, and industry will be greatly attracted to advanced communications channels and the low-cost wideband necessary to link the PBX exchanges, the automated offices, and the databases. The same argument can be made concerning the television-hungry public.

The way is technologically open to worldwide direct broadcasts beaming commercial, educational, and entertainment programs to large audiences equipped with receive-only dishes not very different from present-day radio and TV antennas. The point, however, to be retained is that because of tariffs and technology, U.S.-based corporations could dominate the world's largest information services and, with it, industrial, commercial, and banking activity.

Chapter 14

MANAGING
SATELLITE SYSTEMS

INTRODUCTION

Chapters 12 and 13 properly emphasized the fact that satellites change the whole scenario for world communications and alter our design philosophy. Moving data via terrestrial networks originally designed for voice traffic has been a slow and expensive proposition. The maximum speed at which it is practical to transmit data over a voice-grade circuit is 9600 bps. High speeds are more easily and less expensively achieved via satellite transmission.

In the general case, satellite communications systems involve the communications satellites themselves, earth stations, access terminals, and terrestrial interconnections among them needed to provide a telecommunications service to the user. Three kinds of systems can be formed: point-to-point, narrowcasting, and broadcasting.

Telephone communications is an example of point-to-point service. Business satellites are well established for providing this kind of service. Satellite broadcasting, in which a single source uses a satellite to transmit simultaneously to many earth stations distributed over large geographic areas, exemplifies the other two types of service; both relate to the need of a systems architecture.

Such architecture involves systems engineering, pure economics, market strategy, and regulatory constraints. The analysis of service needs will identify total traffic, its geographical distribution, and the number and nature of competing systems. End-use statistics influence the number, channel capacity, and preferred locations of earth stations in a system. The distribution and capacity of earth stations leads to the network architecture:

- Number, sizes, coverages, and distribution of satellite antenna beams
- Channels allocated to each beam
- Modulation and multiplexing schemes needed to handle the traffic within each beam
- Distribution strategy among the users

Constraints imposed by allocations of orbital arc segments and spectrum influence the size and locations of antenna beams, antenna characteristics, and modulation schemes. Technical regulations for sharing the orbit/spectrum resource impose certain requirements on the technologies available to the system architect for protection from interference.

A major consideration is that of the operational use and the costs associated with it. The systems architecture will have to include considerations of terrestrial interconnection. The larger earth stations will continue to be strategically placed to interface with computers for message multiplexing and distribution, but the proliferation of smaller terminals for industrial, private line, transmission, and broadcast users could create new network architectures.

Should systems planning studies indicate both need and economic justification, the systems architecture and problems of sharing might include cross-orbit beams, satellite-to-satellite tracking, and 60 to 90 GHz, or optical-communications technology, techniques that may come into play for the first time in the 1990s. Trade-off between overall per channel system costs and the available architecture options can lead to an acceptable systems architecture through an iterative process.

At least some of the planning premises should concentrate on the spacecraft channels, combining most service needs for education, teleculture, health care, and a variety of business applications, including electronic mail.

Transmitting at 56 kbps, a computer disk containing 10^8 bits of information consumes 33 minutes; via a 6.3-Mbps satellite transmission, the same disk is transmitted in less than 20 seconds. A large database which currently takes almost seven days to transmit over a 56-kbps land network can be transmitted in only 90 minutes via a 6.3-Mbps satellite link.

Design-wise, this should be achieved through an integrated structure. As I will never tire of repeating, text, data, voice, facsimile, and images are no longer separate entities. Integrated communications is the underlying theme: all these kinds of information move over the same channel. Now that communication forms other than data are first digitized before traveling any distance, data commonly become indistinguishable while en route.

When we talk of managing a satellite system, we invariably make reference to the end-use perspective, although in a more limited sense, this reference tends to be restricted to the basic component parts of the communications network:

- The launching vehicle
- The satellite unit
- The space communications discipline
- The earth station
- The digitization of voice
- System management prerequisites
- Facilities management (operations, supplies, and factory)

Economic factors weigh on decisions whether we take the broader architectural approach or the more limited view. Figure 14-1 presents a "just notice difference" (JND) scale to investment cost per satellite voice circuit per year, under current and developing technology. Al-

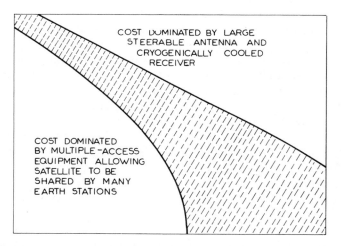

FIGURE 14-1 An order of magnitude (just notice difference) scale to investment cost per satellite voice circuit per year, under current and developing technology.

though the cost of both the satellite and the earth stations is falling for a given level of communications capacity, more significantly the cost per channel with private satellite antennas will vary with the number of channels involved.

EARTH STATIONS

The economics of satellite transmission are governed by the technology involved in satellites and associated devices, the launch mechanism (which we have considered and, at least at present, is outside the control of the telecommunications company), FCC regulations (for instance, satellites no more apart than 3°), differences in signaling protocols (there are about 400 of them now); and existing cable plants (which are not really adequate).

Earth stations will provide the physical link between a business and its satellite network. These are antennas that transmit and receive information from satellites via radio waves. Earth stations can be as small and inexpensive as the $2000, 3-meter dishes used for cable television broadcasts or as large and costly as the $2 million, 30-meter dishes used by major communications carriers, but $200,000 to $300,000 might be the "typical" business earth station cost for some years until the economies of mass change the cost perspectives.

Looking at the systems functions and organization relationships revolving around the earth-bound facilities, we distinguish:

1. *The earth station:* Its functions include: receive/transmit, locate faults, replace field-replaceable units (FRUs), calibrate and align, and checkout/verify operation.

2. *The systems management facility:* Its goals are to monitor and control, provide system and problem determination, exercise system diagnostics, assure performance of remote tests, and maintain losses and statistics.

3. *The maintenance center:* Responsible for repairing and testing FRUs, shipping failed assemblies/subassemblies to supply, performing repair and overhaul, supplying systems, analyzing failure, and maintaining spares. Backing up a network of maintenance centers will be a factory facility able to assure overhaul (beyond the capability of the local center) and to provide engineering services.

The idea behind this system maintenance and management network is that earth stations will be operated unattended, and maintenance (both routine and unscheduled) will be accomplished from the centers we described. Each center will support earth stations in the same

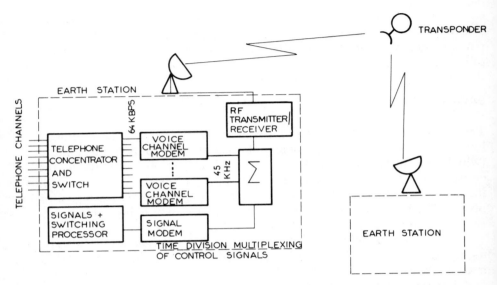

FIGURE 14-2   Block diagram of the single channel per carrier, assignment by demand (SPADE) FDM system.

general geographic area, and local facilities will be supported by higher-echelon maintenance groups with emphasis on reducing personnel costs.

Eventually, systems management will become more complex; as engineering capabilities to manage satellite communications evolve, the satellites become more sophisticated and the number of earth stations increases.

As an example of current technology, Figure 14-2 presents the block diagram of the SPADE (*S*imple channel *P*er *C*arrier, *A*ssignment by *D*emand *E*quipment) FDM (frequency-division multiplexing) demand assignment system. One current estimate is for 4000 stations sending messages to one another as a first step, but there is talk of 40,000 or more stations and the current estimate may go up two orders of magnitude. As the number of satellites increases, the functions to be performed by earth stations will get more sophisticated. Among the technical problems to be solved are those relating to the one-to-many connectivity and to full connectivity.

The principal components of an earth station are the following.

Satellite Communications Controller

The satellite communications controller (SCC) is a highly integrated hardware/software device comprised of processors, storage units, and control programs. It performs the essential TDMA (time-division multiple access), A/D (analog/digital) conversion for voice-grade signals), switching, and other control and processing functions.

The SCC functions as a:

- Time-division switch
- Information processor
- Controller

It also performs forward error correction coding/decoding, voice activity compression (VAC) to improve efficiency of available power, signaling, call processing, circuit switching, echo suppression, formatting, framing, and synchronizing. A major activity of the SCC is multiplexing, and with it (1) multiple access control (satellite access determination), and (2) demand assignment (DA) control for priority determination.

The earth station must multiplex signals received from lower levels for high-speed channel capacity (Figure 14-3). Because satellite communications is a steady process of multiplexing and demultiplexing, these functions will increasingly become embedded in a front-end computer device as protocols, until eventually every computing center will possess its own earth station on the rooftop (Figure 14-4).

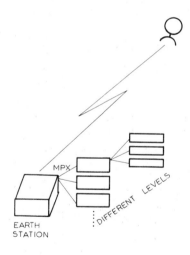

MPX SIGNALS FOR HIGH-SPEED CHANNEL CAPACITY.

IN FACT, THERE IS A CONTINUOUS PROCESS OF MPX, DE-MPX.

FIGURE 14-3 Multiplexing/demultiplexing signals at the earth station for high-speed channel capacity.

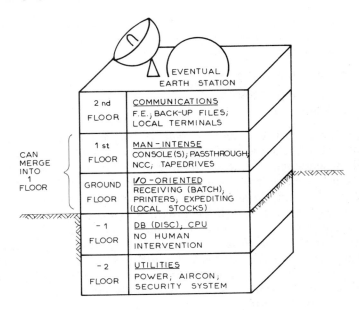

AT THE CONTROL AREA, TO BE LOCATED
AT THE GATE ENTRY, SHOULD BE
INTEGRATED THE CLOSED CIRCUIT TV
FOR SECURITY AND UTILITY AREAS;
ALSO THE CENTRALIZED ALARM
SYSTEM.

FIGURE 14-4   Possible standard structural solution to computer center organization for the 1980s.

In simple words, the controller transforms into a "homogeneous bit stream" the very different kinds of information a user sends:

- Voice
- Data
- Facsimile image
- Video

It converts the telephone's analog signal into digital data like those normally produced by a computer, and it compresses the flow of bits: packing parts of one telephone call into the pauses of another, and squeezing other types of low- or medium-speed data into high-speed droplets.

In principle, the total capacity allocated to a customer network will be divided among the earth stations in that network in accordance with the demand assignment algorithm stored in the SCC. All transmissions via the system will be in digital form.

The division of capacity will be done automatically in response to real-time demand at each earth station. A pooled capacity feature will also be available for use when the customer's traffic demand exceeds a predetermined kbps rate (448 kbps with SBS).

TDMA Burst Modem

The TDMA burst modem performs the modulation, demodulation, and associated functions that enable bursts of digital information to be transmitted through each satellite communications channel on a time-shared basis.

In summary, the functions of the burst modem include modulation and demodulation; generation of the preamble for each burst; overlay and removal, upon demodulation, of an energy dispersion sequence (to reduce adjacent channel interference levels); codeword detection; acquisition; and synchronization with each burst.

RF Terminal

The goal of the RF terminal is to provide the radio-frequency transit and receive functions and to assure the frequency translations between the 14/12-GHz transmit and receive frequencies and the 70-MHz interface with the TDMA burst modem. (This way, TDMA works at 70 MHz while RF transmit/receive functions at 12 and 14 GHz.)

Monitor and Command Loop

The monitor and command loop enables the remote system management facility (SMF) to determine the status and health of earth station equipment and to issue diagnostic and corrective commands. The monitor and command loop is the executive section of the earth station and can itself be monitored through selected, network-wide locations.

SYSTEM RESPONSIBILITIES

To prove the kind of services that are needed, SBS compiled a list of objectives the future system should be designed to realize. They include fully switched, private networks; a system that would interchangeably handle voice, data, text, and image traffic; and approaches able to minimize the need for extensive terrestrial links, allowing the user to have access to data rates substantially higher than the presently available terrestrial service.

Such systems should avoid interference with terrestrial microwave networks, permit economies in space segment use, allow customers to move from today's voice and low-speed data into entirely new high-data rate applications, and vary the data rate on demand, obtaining overflow capacity as required.

System management functions will typically include the following.

System Management Facility

The SMF is basically a data processing facility that maintains cognizance of the status of all earth stations and other systems components, collects traffic data and billing information, and provides support to the maintenance organization via automated system diagnostics.

The SMF should be designed to permit controlled changes to the systems configuration and assignment of resources and to furnish the needed administrative support functions. It should be able to intervene when malfunctioning or performance degradation occurs and additional services or network nodes are to be added.

The SMF can obtain status and health information for each earth station by polling each operating post via multidrop satellite communications channel function, communicating with all earth stations via a satellite link. Among its goals are the collection of system performance and traffic data, the allocation of satellite capacity pool, possible changes in system configuration, and the billing for various services.

RF Spectrum Monitor

The RF spectrum monitor is a sensitive and highly instrumented earth station that monitors the communications signals being transmitted and the timing of information bursts.

Maintenance Centers

Typically, maintenance centers will be distributed geographically in accordance with the location of earth stations, with each center having responsibility for repairs and preventive maintenance for 10 to 20 earth stations.

Effective systems management needs a window into the customer's communications network to monitor network status and performance, change traffic handling priorities, plan the needed evolution in basic services, and collect usage statistics.

Space Segment

Space segment is another vital responsibility. Typically, it includes the satellites, telemetry, tracking and command, the satellite control activity, and the protocols to be observed. If the satellite is 100 Mbps and the packet is 1000 bytes (10,000 bits) the transmission will involve

$$\frac{10^7}{10^4} = 10^3 \text{ simultaneous packets}$$

To achieve this, we must develop theoretical bands of, say, 10 kbps; superimpose many theoretical bands, each requiring a fraction of a second; and establish the limits through a point-to-point handshake. This means significant overhead if new methodologies are not developed. Because the satellite to earth and back delay is 475 ms (¼ second each way), in case we send acknowledgment we bring up delay to nearly 1 second (although there is a company manufacturing a black box, a "turnaround eliminator," which buffers this turnaround).

- Data become pipelined.
- Ack is achieved for the data that are at the bottom of the pipeline.
- This essentially means a very large buffer.

We can also use procedural ways to reduce overhead: for instance, transmitting the same message three times and letting the receiver sort it out.

THE INNOVATIVE SERVICES

Both SBS and its competitors will offer innovative services able to handle a mix of a customer's intracompany requirements, including document distribution, high-speed fax for correspondence, various types of video for teleconferencing, advanced data processing-related applications, and the enhancement of tomorrow's voice traffic.

Several factors favor teleconferencing and make it likely that executives in organizations with widely scattered facilities will use this approach over face-to-face meetings: the burdens and increasing costs of travel; inhibitions on travel, resulting from unpredictable crises; and a growing exposure to education by TV.

With the terrestrial circuits, frame-freeze is a bargain. Since it is offered piggyback on an existing channel that must be justified for its primary purpose, transmission of voice and data, there also exists interest in full-motion TV conferencing. Once an executive has been exposed to teleconferencing, it has an irresistible attraction because of its relatively easy use. Satellites can significantly enhance live TV for teleconferencing and open new possibilities in management control.

As an example of dedicated system applications, a maritime satellite, MARISAT, is designed specifically to provide rapid and reliable telecommunications services to the U.S. Navy and to the commercial shipping and offshore industries. Developed by Comsat, these satellites represent the most significant advance in maritime communications since the advent of wireless telegraphy at the turn of the century. Ships and offshore facilities at sea can be linked instantly via MARISAT with shore points anywhere in the world, and the service includes telephone, telex, facsimile, and data communications.

MARISAT communications are beamed from a shore station to a 720-pound satellite spinning in synchronous orbit 22,240 miles (35,570 km) above the equator, where the signals are amplified and retransmitted on a different frequency to a ship or offshore facility. Ship-to-shore communications follow the reverse path. Each of the three MARISAT satellites in orbit can cover an area of more than 60 million square miles, about one-third of the earth's surface.

Each MARISAT satellite has a five-year design life and operates at three different frequencies to serve distinctly different needs: UHF is for Navy service. More than 400 Navy ships now communicate via MARISAT. The International Telecommunications Union figures indicate that more than 12 million marine telegrams are transmitted ship to shore and shore to ship in a year. Comsat offerings via MARISAT include:

- *Voice:* high-quality two-way telephone fully interconnected with the worldwide telephone network

- *Telex:* interconnected with the worldwide 50-baud teleprinter network

- *Facsimile:* using the voice circuit, facsimiles of manifests, drawings, daily reports, weather maps, and other graphic materials can be exchanged ship to shore and shore to ship

Comsat maintains constant surveillance of the satellites in orbit, monitors their performance, and assures that they are maintained on station. Tracking, telemetry, command, and monitoring services are provided through earth stations at Southbury, Connecticut, and Santa Paula, California, which are linked with the company's System Control Center in Washington, D.C.

Telemetry data received from the satellites by the Southbury and Santa Paula stations are continuously fed through land links into the System Control Center to provide information on the satellites (temperature inside the spacecraft, spin rate, battery voltage). Ranging and tracking data received by the center are used to determine the satellites' orbital positions. Based on analyses of data, engineers in the center issue commands to activate electronic components in the satellites or maneuver them in orbit for stationkeeping purposes.

For the public, the Broadcast Satellite Service (BSS) is viewed by the ITU as providing two kinds of signals: those that can be locally redistributed after reception, and those that are used only by the user at the receiver.

- For redistribution, signal quality at the receiver terminals should be appreciably better than for so-called "direct reception." Such a signal may be characterized as "community reception" and is the type of service envisioned in the United States.

- Other countries, such as Japan, Canada, and India, are planning for direct reception broadcasting, at least in some of their territories.

Microwave use is the common characteristic of both solutions; it will have a definite impact on telephony. Although microwaves have not been widely applied for local communications until recently, where telephone lines dominate, costs are falling while local markets such as cable TV and data communications are taking off at a significant growth rate.

MCI Telecommunications, Southern Pacific Communications, and ITT transmit long-distance calls through microwave or satellite systems at rates as low as 50% of those we usually pay. For $10 monthly (for 24-hour calling) or $5 (for calls between 5 P.M. and 8 A.M.), the user gets a billing code number. He or she then dials the local number connected to the service's computer, enters the code number on the pushbutton dial, and follows with the desired area code and phone number. The user pays 9 cents to 13 cents per minute for evening calls, plus a 10-cent connection charge and AT&T's charge for the local call.

Technology always helps. The development of effective microwave solid-state techniques lends itself well to quantity production techniques and makes it possible to contemplate with confidence a generation of 12-GHz receivers. The most demanding requirement is a low-cost, reasonably stable, and spectrally pure microwave source to provide the local oscillator, but progress has been made in the development of microwave transistors, such as the gallium arsenide (GaAs) field-effect devices.

Savings are significant, even with terrestrial microwave systems. A 10-minute New York–Houston call at 6:30 P.M. costs $2.59 via Bell and $1.39 via MCI. Where AT&T charges for a full minute even if we talk for only part of one, the low-rate services collect on a half-minute basis.

There is more to come in basic economies. It has been projected that by the mid-1980s, when the broadband satellite systems are in full swing costs may fall to between $\frac{1}{40}$ and $\frac{1}{50}$ of today's rates.

THE STEADY EVOLUTION

To better appreciate the coming world of communications, let's take a look at the historical development of the new communications services:

- The first switched digital network was installed by Datran. Financially, the effort was ill-fated, but technologically it led to a new concept: the terrestrial microwave link.

- Packet-switching capability for low-volume data users has been provided for years by Telenet and Tymnet.

- Advanced high-speed data and facsimile services have been added to the services supported by the satellite carriers.

- Networks to link incompatible computers have been planned by several companies: from AT&T's ACS to the ITT Domestic Transmission System.

- The TV/CATV program distribution has been enhanced by taking advantage of satellites (there are now more than 100 receive-only earth stations approved for CATV/Pay TV program distribution services.)

Early in 1977, the FCC authorized the use of receive-only earth stations with small-diameter (4.5 meters or less) parabolic aerials. Considerable use is already being made of audio-distribution via satellite, with 8- and 15-kHz channels. A prototype small aerial system (3 meters in diameter) was developed for Western Union by Hughes Aircraft Company, and a number of cable television stations are already making use of domestic satellites. Rockwell-Collins contracted to provide some 150 earth terminals for the Public Broadcasting System, with 10 parabolic antennas.

Private earth stations will be an extension of this concept and will also use experience acquired with large-scale PBXs. They can be located anywhere a customer wants—on the roof of an office building or in a parking lot. Employing little-used high frequencies that escape interference from conventional microwaves, the satellite system can provide customers with the same high-quality service in Manhattan as in the Nevada desert, and there will be no waiting for AT&T to lay on the right line.

At a $200,000 average production cost the earth station will be cheap enough for a corporation to lease a dozen or two, and forge direct celestial links among its offices and plants scattered about the country or the world. And whereas standard satellite communication today requires the leasing of separate channels to carry a broad range of voice, data, graphic, and video information, the new system will be able to carry all these on one channel which can be adjusted to a user's changing routes, traffic density, and types of messages.

Earth stations will interface with interconnecting facilities to customer-provided PBXs, foreign exchange lines, data terminals, and other equipment. Units that are located on the same customer premises as the earth station may be connected via plant cabling. Locations remote from earth stations will be interconnected via conventional terrestrial facilities such as those available from the terrestrial common carriers.

The earth station will provide the necessary interface adapters to ensure electrical and supervisory compatibility with the interconnecting facilities. Where needed, service between on-net and off-net customer locations will be provided. Voice and other analog signals will be converted to a digital format at the originating earth station and converted back to analog signals at each terminating earth station.

The user of network facilities will have available an assortment of advanced features, including:

1. Flexible voice and data conference arrangements
2. Multipoint distribution of text and data, which can include document distribution
3. Teleconferencing, including multipoint video conferences
4. Hot line and other priority connecting features
5. Network access control features

What is different about the SBS and similar advanced technological ventures is the linking of many innovations into a single system. Sales representatives could use video facilities to deal with distant clients, saving travel expenses and spreading their talents among a greater number of prospects. By using slow-scan video conferencing in which the picture changes infrequently, one potential customer figures that it could leave at home some 280 to 300 people who must now be brought to headquarters for monthly two-day management meetings.

Texaco sees so many attractive possibilities that it is thinking about restructuring its entire organization to take advantage of them. Large corporations and government agencies are interested in the progress of SBS-type ventures because of potential cost savings. Many of these large users spend millions of dollars annually to lease communications facilities, and the basic economies of the new systems will permit very considerable savings.

System flexibility makes for greater use of capacity and lowers transmission costs. The satellite's capacity will permit a company to tighten its control over remote operations by increasing the amount and timeliness of the text and data it can gather on inventories, sales, production, and so on. The consequences for efficiency, organizational structure, and management style could transform the way we do business.

Chapter 15

A ROLE
FOR COMPUNICATIONS

INTRODUCTION

"Compunications," the merger of computers and communications, is one of the vital links in modern business and can be put into three principal types of use:

1. Replace lengthy telephone calls.
2. Substitute for short memos.
3. Introduce new communications methods not previously possible.

But the conversion of old, time-honored approaches and the introduction of new solutions necessarily involves setting goals, performing feasibility studies, projecting architectures, proceeding with systems analysis, and placing emphasis on user training. Since we know from other experiences that the new media alter the classical communications procedures and ways to do things, we must create new images.

That is what the Department of Agriculture's Research Division has done by instituting "Project Greenthumb." Another example is the Videotex trial begun in Washington, D.C. This test will eventually involve 64 sites, including offices, homes, and educational institutions, with information services coming from *The Washington Post*, thε Smithsonian Institution, District of Columbia libraries, and major U.S government agencies.

This and similar processes currently going on in the United States are important in handling sales orders, confirmations, inventory infor mation, orders to suppliers, and daily registration of any type. Among its advantages are the use of the public highly distributed telephonε network, reduced decentralization of physical records, and immediatε connection to central resources.

American manufacturers are speeding up their announcements o: two-way interactive graphics systems. Seven firms demonstrated sucl systems at the 1980 National Cable Television Association show. Witl recent rulings stating that cable companies must provide serviceι beyond broadcast, there is an even greater demand for high-quality interactive two-way capabilities for business and home use.

The market will certainly be quite competitive. For $399, con sumers will be able to purchase from Tandy's Radio Shack stores a videotex terminal that will link any TV set to the "viewdatabase" such as the central computers operated by CompuServe Inc. in Columbus, Ohio. In early 1981, General Telephone & Electronics Corp. started marketing a competing videotex service over its Telenet data communi cations network, and AT&T had already invited other companies to combine resources to create and market viewdatabases using the phone lines and home terminals.

AT&T has undertaken a variety of experimental Videotex applica tions that include classified, display, and catalog advertising. "Ma Bell" considers it a natural extension of its existing directory information services (white pages and Yellow Pages) and says that some newspaper publishers are trying to freeze the telephone network out of operations made possible by the technological revolution of computers. The infor mation services are to be provided by the unregulated subsidiary that the Federal Communications Commission (FCC) has said it should set up and with which newspapers and other information providers could compete on equal terms.

A different way of making this statement is that Electronic Yellow Pages are an alternative to the directory books published every year. They can be updated daily, so they can offer the timeliness that the annual directories cannot. They carry advertising, as do the printed directories, and that poses a problem.

The problem is not for the telcos which stand to gain advertising revenues, but for the newspapers, the traditional suppliers of timely information, including advertising. No wonder the American News paper Publishers Association (ANPA) opposes this move—while surpris ingly nobody asks the consumer public's opinion on the matter.

For instance, in one of the experiments, conducted in Austin, Texas, the publishers, who had $14.5 billion in advertising revenues last year, say the test is an example of the kind of "unfair advantage" that they contend American Telephone & Telegraph will have in developing services directly competitive with newspapers and other publications. The wheels of the compunications engine have started turning.

THE TELEPHONE COMPANIES TEST THE MARKET

The telcos experimenting with electronic phone books typically provide free computer terminals on separate lines to a sample of private homes and 60 businesses. A user could consult advertising by real estate agents, supermarkets, department stores, and other businesses as well as listings by brand name. Recent experiments go substantially beyond earlier AT&T tests in New York City and Albany, New York, which generally gave users access only to the telephone listings. The object is to gather as much information as possible about what customers want and would use in a market potentially worth billions of dollars yearly.

Statistics help document how radical is going to be the coming change due to the information technologies. The French PTT first launched an electronic phone book on videoscreens. When completed, by the latter part of this decade, this will save about $100 million paid annually by the PTT for printing and distributing the traditional, paperbound telephone directories. As a result, 40% of the yearly work done by the French national publishing house will be eliminated, together with 3500 jobs directly related to the French telephone system's inquiry (directory) service. On the other hand, this will lead to an impressive internal market of 32 million telephone video sets. Such a system would represent an internal market of over 200 million telephone video sets in the United States, double that amount worldwide.

This development comes none too soon, but still in time to stem the mounting stacks of paper. Let us always remember that paper does not flow like water; it accumulates like garbage. In France, the telephone authority stated that the consumption of paper for telephone books increases at the square (power) of the number of subscribers. There are 14 million principal line connections today, each requiring a telephone directory. There will be an estimated 30 million in 10 years.

To achieve cost savings and modernize its system, the French telephone authority is prepared to equip the telephone sets with a video set at no cost to the user. Through an easy menu selection access, it will be possible for any subscriber to obtain response to his or her inquiry about telephone numbers—the text and database being updated steadily and readily accessible to any request—as contrasted to the voluminous phone book, which, at best, is updated yearly.

But is the service worth the cost? The French PTT made an experiment in Saint Malo which demonstrated that the technology is ready and available. The questions are: How much will the electronic phone book cost? How much will it offer? Are users ready for this service? The cost will not be trivial—and in France one hears the term "a second Concorde" (the technologically advanced but commercially ill-fated Anglo-French supersonic jet aircraft).

The PTT and Matra are the moving gear behind this process, eyeing a market (if it ever develops) that might reach the stated 32 million video sets in France alone. Still the question remains: Will it pay? To find the answer, Continental Telephone (to which we already made reference) conducted a real-life test together with the French at Big Bear, a mountain resort near Los Angeles. The main emphasis is placed on the Yellow Pages. Besides the name, address, and phone number, the advertiser will be able to buy space, up to three lines. The experiment seeks to determine whether the business community will adopt the service and whether the ads will pay the cost to make the video process free to the consumer.

Cost does not seem to be a French preoccupation, where "it's up to the government to decide"; therefore, the results of the American trial are more significative than the new French experiment scheduled at Ille-et-Vilaine, south of Paris. In the latter, the PTT will be installing some 250,000 video units "free of cost" to subscribers, but the government will have to pay 1000 to 1200 francs per set (about $150), with the taxpayer footing the bill until further notice.

If cost and benefit are two of the unknowns, the other is the normalization of the technology. On this, no one seems to be in a hurry. GTE's recent decision to discard the Prestel technology (for which it paid a large amount of money last year for U.S. marketing rights) in favor of Canada's Telidon is a major setback for Prestel. No wonder that both Prestel and Antiope researchers are now working on a dynamic redefinable character set to provide a Telidon-like quality for their systems through Alpha-DCRS (dynamic character redefinition set). This capability, however, is not expected to be available before 1982. Videotex technology, the blending of telephone and TV, holds the greatest premise for person-to-person communications.

Looking at this issue in a historical and technological context, we can see that prior to the fifteenth-century invention of movable type, text distribution was highly constrained by the slow speed at which original manuscripts could be copied. Instead, the reader traveled great distances to the repository of the original document. Printed books changed the perspective, allowing the text to travel to the reader, but once printed, the contents of the book are static.

Online inquiry, updated on a "real enough time" basis, is always current, actual, and able to be called up instantaneously in videocolor. The capabilities this field opens up can greatly enhance the profitability of a telephone utility which currently earns a mere 5% of annual revenues from datacomm. Interactive videotex also involves a high proportion of intercity communication compared with conventional telephony.

FIVE AREAS OF APPLICATIONS

Careful study is a prerequisite. Prior to engaging in the time-consuming and usually costly process of additions, deletions, and alterations to compunications systems, we should look carefully at the functions we want the evolving compunications systems to perform. The phases such a process involves are fairly well defined and center around message:

- Creation (including editing)
- Distribution (to one or more recipients)
- Reception (including error detection and correction)
- Annotations
- Time stamps

Annotations and time stamps are new concepts in electronic message environments. The latter is needed for systems housekeeping. The former is viewed as a basic requirement for asynchronous-type solutions, enabling coauthorship of messages, articles, and books, but also conferencing, project tracking, and other party-line information requirements.

Quite understandably, different degrees of sophistication can be advanced. In line of increasing complexity we can distinguish:

1. *Document distribution* systems projected for either or both goals: transmission of non-computer-generated paperwork; high-volume document exchange

Facsimile, online plotters, and intelligent copiers are examples of modern approaches to document distribution, an elder reference being teleprinters. Facsimile units scan printed, pictorial, and graphic documents and as the resolution improves (from the original 96 × 96 dots per square inch to today's 200 × 200 and the projected 300 × 300 dots), drawings, pictures, and character sets can be transmitted with good quality.

Facsimile speeds are a crucial factor: they have increased from 6 minutes per page to 1 minute per page, and a further improvement is projected to between 1 and 5 seconds per page. Intelligent remote copiers have more capabilities than fax but also cost more. IBM's 6670 is an example.

2. *Real Enough Time (RET) communications*, where a large number of users can participate. These may be of the broadcasting/ narrowcasting variety, in which case it is possible to assimilate them to the preceding class, or of an interactive nature best served through the user-friendly approaches of interactive videotex (viewdata).

Broadcasting videotex (teletext) uses the blanking intervals of television broadcasting to transmit text and data frames, sending 1000 characters at a time between each pair of television picture frames. Teletext transmission can be encapsulated so that only sets with a decoder attached may project the information. Acceptable response time, however, limits the number of frames to be handled, early experiments having stored some 300 frames in the database of broadcasting videotex.

Interactive videotex has no limitations of this kind, and the computer-supported database may be dimensioned to user needs, with each user becoming an information provider if he or she chooses. More important, the telephone system viewdata uses are universal, with 360,000,000 subscribers available online worldwide.

3. Computer-supported creation/reception of message services, which include distribution but also more complex editing and retrieval capabilities than those of interactive videotex

Computer-supported message systems find in electronic mail their best expression in terms of present-day services. Such capabilities are offered by value-added networks, both current and projected. Office Automation promises, however, to promote a broader range of services by making all workposts able to communicate with one another in addition to performing data and text-editing functions. (There are about half a million word processing systems installed around the world, at least 50% being in the United States. However, the large majority of these units do not have a communications capability and cannot be retrofitted with it. Furthermore, computer-supported message systems require a first-class storage and retrieval capability.)

4. The still ill-defined field cumulatively called *computer conferencing* and including systems widely varying in capabilities: from electronic bulletin boards, to the coauthorship of documents, and asynchronous teleconferencing

Few computer conferencing systems are currently available, but judging from the predominant thinking in this field, we can say that their technical characteristics call for active participation of the recipient, as contrasted to the rather passive solution of storing incoming messages on a database and waiting for the station they are destined to access. A variety of frills may be added to this reference, such as "whispering" to selected conferees rather than sending messages viewed by parties in preestablished distribution lists.

Memos, announcements, and other types of news can be supported by such systems, which also have coauthorship capabilities, with each level of higher sophistication involving thorough organizational changes and steady training perspectives.

The compunications classes that we have examined so far are basically *asynchronous*:* dissociating sender from receiver through the use of the appropriate protocols and a store and forward facility. Asynchronous-type communications have been used for data purposes and, to a lesser extent, for text, but this concept is now extended to a full range of activities, including data, text, image, and voice.

By dissociating the sender from the receiver, asynchronous-type communications offer the user significant degrees of freedom in designing and implementing his or her system. There are also, however, cases where this approach cannot satisfy requirements. The alternative is a *synchronous* solution where sender and receiver are online at the same time, a typical example being a telephone call.

5. *Synchronous teleconferencing*, implying simultaneous use of facilities, with all participating points in the system being present and equipped for send/receive

Teleconferencing (TC) has been experimentally proved in early 1978 in an experience which involved three users at different sites and time frames: North American Rockwell, Montgomery Ward, and Texaco. IBM was the system manager; Comsat provided the satellites, Hewlett-Packard the computers, and NEC the frame-freeze equipment. (This experiment is explained in the following section.)

Applications are just beginning. For example, in Los Angeles, Atlantic Richfield is designing a complete teleconferencing network so that key company employees can confer with one another visually by use of satellite hookup and wall-sized projection screens. With this network in operation, executives in Philadelphia or Dallas will not have to fly to Los Angeles for their regular weekly meeting. Instead, they

---

* Not to be confused with the asynchronous, start/stop type protocol in data communications. Distributed information systems basically work on an asynchronous basis by decoupling different processes running in parallel on the network—thus avoiding delays, contention, and interlock—even if they may use synchronous datacomm protocols.

will walk to a room equipped for a teleconference. The system, part of a $20 million company-wide communications effort, is expected to save ARCO $50 to $60 million annually in travel costs.

Teleconferencing can take place with or without image. Typical users are industrial companies and financial institutions with regional, national, and international spread. A number of goals can be stated. The two most important are:

1. Permit several persons at different organization levels to communicate without travel.

2. Transfer all arguments and decisions company-wide in an instantaneous way.

The structure of a teleconferencing system can be hierarchical or cascade, with different levels of control (Figures 15-1 and 15-2). It is possible to insert registration means as well as to provide other services. Using the same line, we can have freeze-frame (slow-motion television) and facsimile, also telegraph channels, long-distance slide transmission, and so on. Typically, the line will be high quality, full duplex. Teleconferencing systems will work multipoint and can also perform selective services.

Teleconferencing service is most inexpensive when the user has private lines at his or her disposition. In this case, the basic difference is upgrading. Of all new services, this is perhaps the easiest to install (because companies already have private and FX lines) and, with some minor exceptions, the technology seems to be there. Yet TC systems have still to provide great excitement. Companies feel that by far the first priority should be document distribution, followed by inter-

HIERARCHICAL

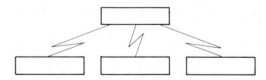

OR CASCADE

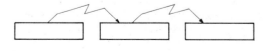

WITH DIFFERENT LEVELS OF CONTROL (MASTER)

FIGURE 15-1 Possible structure of a teleconferencing system; two alternatives are presented: hierarchical and cascade, with different levels of control.

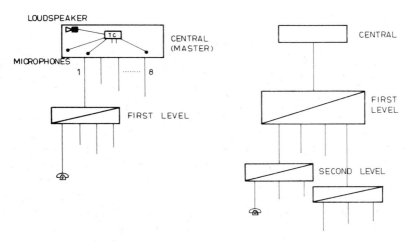

FIGURE 15-2   Hierarchical levels of a teleconferencing network and equipment use.

active videotex and the other stepping stones in terms of a greater sophistication.

SLOW-SCAN TELEVISION

The pivot point of the new compunications services is the universal telephone network, reaching some 400,000,000 ports. This network is steadily modernized, tremendously expanded, and often converted in the basic services it supports.

Telephony as we know it today has a great many assets. The facilities that the telephone network support necessarily reflect the latter; we can state as examples: freeze-frame viewdata, and facsimile. A freeze-frame system (FFS) is also known as phone line television (PLTV) and slow-scan television (SSTV). The goal is to use the telephone lines to transmit images in a process quite similar to that used by closed-circuit television (CCTV).

Figure 15-3 presents a typical CCTV setup. It has been used in a wide range of applications: area control; visualization of alarm points;

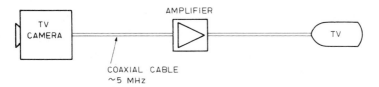

FIGURE 15-3   A typical closed-circuit television installation.

the possibility of transmitting, without modification, text, designs, graphs, signatures, and photographs. It can take place in the most varied environments: with temperatures varying between -50 and $+200°$ C, in an atmosphere full of explosive gas, in the presence of radioactive contamination, and so on. It is a costly setup but a polyvalent one, useful to many situations characterized by stringent requirements.

Within a broad range of applications, we are faced with a wide choice of terminals. The criteria are:

- Quality of image
- Nature of the object (to transmit)
- Operating considerations (including telecommands)

The transmission phases involve handling the video image in the form of an electrical signal and visualizing the information on the monitor in the form of horizontal lines. The image definition is made in a dot matrix: number of lines and points per line. A typical matrix (under current technology) consists of 128×256 points/line, while 256×256 are the new announcements. The time needed to transmit these elements is an equally important characteristic.

The use of telephony to substitute for the high-capacity coaxial cable is a trade-off: universality and the ability to transmit long-haul against a necessary reduction of bandwidth. The latter brings up the need to convert a video signal to voice frequency. Figure 15-4a presents the system configuration: the freeze-frame approach permits the handling of a standard video signal successively transmitting fixed images over a telephone line. The updating time is variable as a function of channel capacity and image definition.

The conversion process is seen in Figure 15-4b. The A/D converter proceeds with the reduction of the video information to the desired number of lines. A storage device assures a rapid scanning capability (at about $1/50$ of a second). This memory is read by the next converter, which assures slow scanning (at roughly 8.5 inches) with signals compatible to the telephone channel. The choice is based on the definition of the image, the maximum time of update, and the nature of the transmission system (as a function of the available line).

The application of a freeze-frame system in a city environment may have different characteristics, given the economic aspect. As in every technicoeconomic solution the goal is: control over cost.

- TV terminals cost less than $1000 and can be used for a variety of purposes, including videotex and entertainment.
- The main cost is transmission, and we must reach a good usage capability.

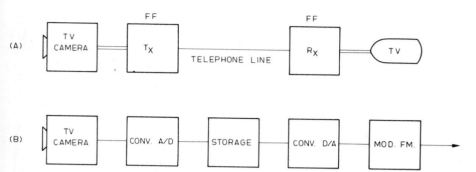

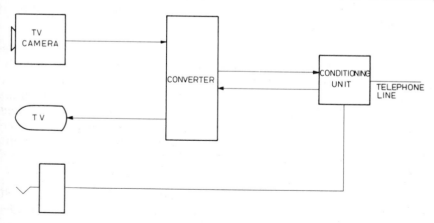

FIGURE 15-4   (a) A system configuration for slow-scan television; (b) the analog-to-digital and digital-to-analog conversion process.

The 64 kbps can be easily achieved in city lines, but for intercity (long-haul) 2.4 kbps is the best line we now have available. (With 64-kbps lines, resolution is 128 × 256 and updating is achieved at 3.1 s.) Transmission types can be distinguished as a function of the telephone line to be used and of the desired facility. The freeze-frame system may work bidirectional (Figure 15-5) with the addition of a conditioning unit for exchanging voice, thus materializing a "videotelephone" capability.

A central alarm system for controlling accesses, installations, and devices is a valid example of FFS application. Typically, with the voice band of 3000 to 4000 Hz for data and alarm, this system will receive and analyze the signals from the periphery. In general, though, the process is good for image, penalizing alphanumeric information because of the amount of useless dots to be transmitted to give the needed data. Machines under study, however, are able to reduce this redundancy.

FIGURE 15-5   A freeze-frame system able to work bidirectionally.

FACSIMILE

Over the telephone lines, facsimile supports hardcopy: at the receiving end it reproduces messages, pictures, and any image sent from the transmitting end. The facsimile phases are:

1. Synchronization (needed to put in phase the sender and receiver)
2. Scanning
3. Transmission
4. Reception
5. Registration

Scanning is the line-by-line exploring of the image. To reproduce effects through an optical-electronic device, either the drum or the flatbed principle is followed. The writing processes at the receiving end can be electrochemical, electrothermal, electromechanical, or electrostatic. Future developments will focus on reducing transmission time, eliminating manual work (paper, insert, start, stop, etc.), increasing quality, using telephone lines in their valleys, reducing the unit cost, and improving reliability.

Basically, however, the process remains unaltered. Since its invention in the nineteenth century, facsimile permits the long-distance reproduction of alphanumeric and graphic information without any need for codification, with graphs, text, and data transmitted and reproduced *as is*. This is properly appreciated by the users, as recent research demonstrated:

- 76% of the users said they were satisfied.
- The others complained about difficulty in connection, quality of transmission, and cost.

The last is not surprising, as 50% of users transmit one page or less per day. Costs drop sharply as use increases (Figure 15-6). Significantly, among users, facsimile services replaced 64% of the traditional post, 41% of the telephone calls, and about 10% of telexes. The primary users of facsimile have been manufacturing companies, but financial institutions and other organizations are also adopting the service. In practically all cases, the type of connection is point to point, with no evidence of multidrop. The terminal technology has been standardized by CCITT into groups:

- *Group 1* is analog, follows Recommendation T2, transmits 180 lines per minute (lpm), and takes about 6 minutes to send a standard page.

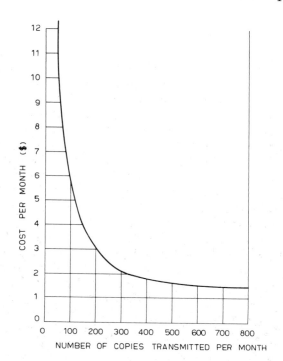

FIGURE 15-6   Indicative costs of facsimile transmission as the approximate function of the number of copies transmitted per month.

- *Group 2*, also analog, follows Recommendation T3, transmits 240 lpm, and requires about 3 minutes.

- *Group 3* is digital. This is fastest, sending a page in 1 minute. By changing resolution it can transmit in 30 seconds. (Also, digital is much faster; Group 4 is under development.)

IBM and other companies advance developments with a totally new technology requiring 3 to 5 seconds for the typical (DIN A4) page, but they are out of the CCITT standard and cost some $35,000 versus the $12,000 to $15,000 for the 1-minute device. (The former machines can also copy the document that arrives.)

In Group 3, lack of CCITT standards has led to terminals that are incompatible. Furthermore, although quality and speed are much better, copies cost more (Figure 15-7). Group 1 devices cost about $2000; Group 2, $4000 to $5000. Transmission costs make the difference. Fast facsimile is used for transmitting newspaper pages. The CCITT standard for networks is T10 for private lines and T10b is for public lines.

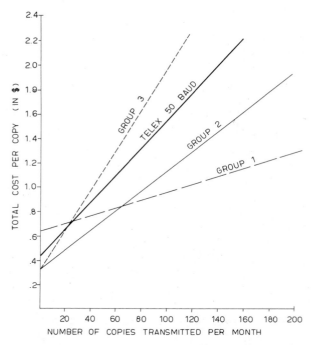

FIGURE 15-7 Costs per copy of facsimile transmission as a function of the number of copies per month.

Chapter 16
NETWORK ARCHITECTURE

INTRODUCTION

Chapter 15 has brought under perspective the changing role of telephony: POTS (the plain old telephone service) is structurally and dynamically changing. The new services outweigh the old in complexity. Together with the rapidly dropping cost of small computers, the communications capability now under development propels corporate users into adopting a different stand toward telephony.

Standalone computers proliferate within a distributed information environment, but then comes the moment when these standalones must be interconnected with the central computer; they must access distributed databases and provide the end user with a facility characterized by network-wide access. This brings the need for a network architecture.

Through the software routines that compose it, a network architecture acts as the traffic signals in regulating the flow in a message system which may carry voice, text, data, images, or a combination of all of them (the preferred solution). This is the essence of "compunications." During the next 10 years, computer-based message systems will have a great impact on the way business is done in our society, greater than the telephone had on business practices during the last 100 years.

As documented by the experience with the first two generations of online systems, applications are the primary factor that determine the type of compunications. We used teleconferencing and facsimile as examples. Other expanding datacomm applications are credit inquiries, electronic funds transfer (EFT), electronic mail, stock brokerage information, travel reservations and scheduling, point-of-sales data collection, sales order entry and inquiry, inventory control and reordering, and production monitoring and control.

The list can be long, and the longer it gets the more polyvalent are the demands posed on the network architecture. A good deal of the online applications are accounting and financial; others concern management information; still others document cataloguing and retrieval. Every sector of the society is part and parcel of such developments, health care, hospital, and other medical information being further examples. Before too long, there will be rather significant home-sector applications as the microprocessor-supported home information systems evolve and cable TV reaches more and more homes.

COMPETITIVE INFORMATION NETWORKS

At the dawn of the computer age, in the 1950s, nobody worried much about getting information from one computer to another. The mainframes were housed in air-conditioned rooms, and operators had to work close by in order to use them. Remote unintelligent terminals in the 1960s, operators were allowed to use the computer from locations at a distance from the computer room. Then, in the 1970s, we experienced distributed data processing based on minicomputers and using intensive data communications.

As the datacomm requirements grew, a host of suppliers were trying to enter the market with products that would permit one computer to talk to another, whether they are across the country or across continents. This involves long-haul networking systems that would be used to link together individual offices wherever they may be.

Banks, merchandising firms, and industrial enterprises may soon have an information outlet in the wall, just as they now have outlets for electricity and voice telephones. The information outlet would be connected to a cable that ties together a local area network equipped with personal computers in executive offices with other information systems, including local, regional, and remote files or duplication services.

Information networks dedicated to the efficient handling of datacomm needs, originally known as Value-Added Carriers and Value-Added Networks, developed during the mid-1970s. Tymnet and Telenet in the United States, Datapac in Canada, PSS in England, and Transpac in France are but a few examples. To a substantial measure, they followed the lead of Arpanet and implemented packet-switching principles. (This is not an exclusive reference. For instance NPDN—the Nordic Public Data Network for the Scandinavian countries—is a modern datacomm oriented carrier, yet it has circuit switching.)

Two types of ownership characterized the VAN launched in the mid-1970s to 1981: public telco authorities (PSS, Transpac, NPDN) or small companies which started on a pioneering road with shaky finances and either found a profit-making basis (for instance, Tymshare/Tymnet) or sold to larger organizations (GTE Telenet).

The year 1982 brought two private heavy weighters into the VAN arena. In February 1982, IBM announced its Information Network, a national data communications system that provides services similar to Tymnet and Telenet. One of its future objectives is to offer computer users the opportunity to unload their entire teleprocessing front-end activities, a business which is at the very heart of IBM's revenue base.

In June, *American Bell*, AT&T's long-awaited unregulated subsidiary, came into the world and brought along its system. The company immediately had 1000 employees and $59 million in assets from its corporate parent. Its first offering is *Advanced Information Systems/ Net 1*.

"The new service will do for data transmission what the switchboard did for telephony," said the chairman of AT&T. American Bell's main competitors are IBM and GTE. IBM's Information Network makes possible high-speed transfer of computer data. GTE's subsidiary Telenet has about 700 customers for a similar system.

Net 1 provides data storage and transmission, programming, and the management of data communications networks. It permits communications among terminals and computers; hence, so far incapable of direct communication. Users will connect their terminals to the AIS/ Net 1 service through a common carrier providing an access line to one of several service points set up nationwide.

At the service point, protocol translation used by different data processing devices takes place, and access to processing and storage is provided. As with the other VAN, a common-carrier packet-switching network is employed to transport data among the service points.

Net 1 has been announced as a public computers and communications network designed to interconnect the majority of terminals and computers in operation today. Its hardware gear is IBM Series/1 minis and DEC VAX computers. The software is projected to help connect on both an intercompany and an intracompany basis.

AT&T has developed a list of intercompany services whose development would help establish new means of doing business. A new term "customer premise equipment" (CPE) has been coined to identify the Net 1 terminal stations at the customer's site. CPE hardware characteristics might range from microprocessor-based units to 32-bit machines.

As a network operation, Net 1 helps reduce many of the headaches and costs involved in running leased lines and private networks. American Bell will contract for a package of network services, which as far as the end user is concerned, will appear as a private communications system dedicated to his needs.

Through its wholly owned subsidiary, AT&T will be hard-selling bundled solutions involving hardware, software, and know-how to satisfy a range of customer requirements. Transitional capabilities are also said to be assured so that large users won't have to discard installed equipment or write complex software to interface different computers.

The way AT&T looks at it, a major competitive issue with Net 1 is that it can provide compatibility among datacomm units over a broad range of transmission:

- Speeds
- Codes
- Line controls

in point-to-point and multipoint communications on both private and public transmission lines. This assures businesses a flexibility to introduce new, more productive data processing, text processing, databasing, and datacomm systems while continuing to use existing investments.

Since the AT&T intention is to develop Net 1 as a grid of computers and communications services, American Bell's offering might replace not only leased lines but also the mainframes, offering standard application programs and giving the customer the ability to tailor them to his needs. (The first available packages were written in COBOL and run under the VMS operating system on VAX computers.)

Some sources are suggesting that market planners and software developers at AT&T have produced ideas for a range of intercompany services that would create an *electronic marketplace*. Through this facility, companies could determine availability of products, make inquiries, and place orders.

To reach a critical mass of customers and ensure profitability, American Bell also seems to be oriented toward a product line able to help widely dispersed firms in keeping track of cash at the end of each working day. Banking, brokerage, and insurance are some of the industry areas at which this marketing policy is oriented. A similar statement can be made of merchandising and several of the professional markets.

THE GOAL OF A NETWORK ARCHITECTURE

A main objective of a network architecture is to assure a task-to-task, job-to-job, and process-to-process communication. Given that the software routines supporting the architecture are written for the general case of implementation, this must be done to a large extent without knowing the network's exact topology and hardware/software characteristics.

The architecture must support device sharing: one machine must be able to run the devices of the other machines or hosts. (A host processor is a standalone computer system able to make use of network facilities as the need arises.) Typically, the distributed (local) intelligence will consist of minicomputers, terminals, and word processors with a simple operating system running in a monoprogramming mode. Devices of many types will be attached to the network, but the fundamental requirements are invariant (Figure 16-1).

Network architectures are based on protocols. Their role is to describe the format of the message being transmitted, including the addresses of the sender and the destination receiver or receivers, the message packet, and various sequences of control and error checking bits or bytes. Without the formalisms supported by the protocols, communications would not have been possible.

Another basic prerequisite of a network architecture is to support *file sharing*, or more precisely, distributed database access. This must be done with a dependable security/protection mechanism; use directory services; and rest on an organizational understructure (to be provided by the user) whereby all information elements constituting the distributed database will be provided with the appropriate text and

NETWORK FUNCTIONS

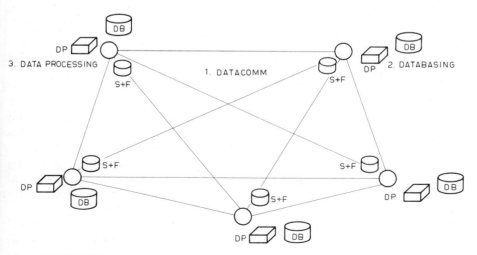

FIGURE 16-1   Fundamental requirements for value-added networks and terminal attachment to them.

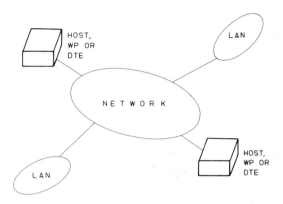

CRITERIA FOR CHOSING NETWORKS ARE:
ERROR PROTECTION; AVAILABLE PROTOCOLS;
GATEWAYS TO OTHER NETWORKS, TRANSMISSION
SPEED; STORE AND FORWARD FACILITIES; AND
PARAMETRICALLY DEFINED FUNCTIONS.

FIGURE 16-2   A basic network function: communicating databases.

data definitions; also the links and controls relating them to the applications programs (processes) and hardware devices authorized to access them. Figure 16-2 identifies the network functions of communicating text and databases.

A third group of primitives (basic functions, imbedded commands) of a network architecture is that of *downline loading, upline dumping,* and *loopbacks.* Such operations must take place in a manner transparent to the user, by using the facilities available in the network: Text, data, eventually voice, programs, and system commands are examples of the entities which need the type of described support.

"Routethrough" and "Passthrough" are two other facilities. The systems architect, the analysts, and the programmers at the user's side do not need to know how the network routes the messages to reach their destination, but they must be assured that the mechanism works and that the network architecture efficiently supports this function.

Passthrough facilities will see to it that the distributed devices— hosts, minis, terminals, and word processors—work without professional operators. What there is in operator action is at the center, the distributed resources being commanded by remote control.

A network architecture supporting the X.25 protocol will be based on the *virtual circuit* concept. A virtual circuit (logical circuit, logical path) is a point-to-point switched or permanent circuit over which data and command packets transit; examples of command packets are reset, interrupt, and flow control. An alternative approach (not embedded in the X.25) is the "Datagram," a section of a message (typically of 256,512 characters) individually routed through a packet-switching network (Figure 16-2).

Not to be confused with virtual circuits are *virtual devices* and *virtual programs* running on a network and sharing resources by creating virtual-level DTE (data-terminating equipment). An example of virtual DTE is given in Figure 16-3. The concept guarantees a homogeneous logical configuration (while the different physical DTE are not homogeneous) so that several programs can have access to spread-out devices *as if* they were local and homogeneous. This provides for optimization of resources and the avoidance of duplications and incompatibilities.

Stated in different terms, a network can have either homogeneous or heterogeneous devices; the former is the rule. But the software routines supporting the network architecture are unique. How can we manage the system? The solution is found in defining virtual DTE; these are able to guarantee basic network-supported functions, keeping as standalone the other services which do not reflect themselves in the network software. This makes feasible designing the network software around the basic functions served which are common to every attached device.

A crucial subject with any network architecture is the ability to perform *online network and device maintenance.* This calls for online tests (hence loopbacks), quality histories, and remote diagnostics. Associated with it is the failsoft capability: components fail, the overall capacity is reduced, but the network is still on. To answer failsoft requirements network-wide, the topology and network management facility should remain dynamic. Networks must be characterized by a 99.99% reliability.

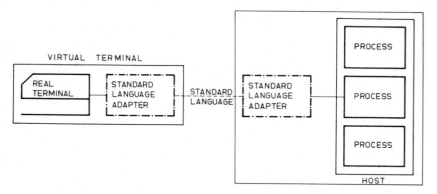

FIGURE 16-3   Example of a virtual terminal capability for network attachment.

Availability is the probability that a system is running at any point during scheduled time. It measures the percentage of time during which the system is operating properly. *Reliability* is the probability that the system will give satisfactory performance for a preestablished period of time when used in a manner and for purposes outlined in advance.

The network management facility is exercised through the Network Control Center (NCC). Through it the tests, remote diagnostics, and quality histories are supervised, and the NCC makes sure that the proper care is taken so that the network in its overall operation is immune to noise, error sources, and other failures. Software routines detect and correct errors and provide for automatic retransmission to transfer data correctly. The user must *not* worry about how this is done.

Accomplishing these functions is an involved, demanding job. The faculties that we have been outlining must be guaranteed by the network architecture in a timely manner (response time). *Network response* involves microscopic control decisions being made, second by second. Their stream conditions the network's response.

With computers and communications systems, *throughput* measures the steady-state work capacity of the system, the time necessary for processing a specified work load. *Turnaround* is a dynamic measure; it indicates the delay between the presentation of input to a system and the receipt of output from it, given the job stream within which it operates.

Prior to adopting a network architecture, the user must be contractually assured that the response time is acceptable: for instance, 95% of transactions happening in less than 2 seconds, the balance not exceeding 6 seconds. Increases in the response time, such as those caused by errors, indicate network degradation.

HANDLING MESSAGES

Networks transmit messages. A *message* is any bit string of data, text, image, or voice that is communicated. As such, messages encompass the entire horizon of information exchange; this may involve financial documents, checks, sales orders, bills, personnel records, advertising, or engineering documents.

Modern communications systems are neutral about the content of information. A network is a transport mechanism and should not be confused (as often happens) with data processing, databasing, and end-user functions. The problem with hierarchical large network structures is that they are too difficult to manage through just one point of control, and too inefficient to route all traffic through.

Messages are carried by a number of diverse communications systems, and communications theory provides one important perspective which should be emphasized strongly: *The channel is far more significant than the device.*

For the most part, the capacity of communications channels limits the speed of communication devices; rarely has the opposite been the case. Efficient solutions to the communications problem have recently been profoundly influenced by the convergence of communications and computer technologies.

Considerations such as these bring message processing from the world of accounting and order entry into the mainstream of business survival. As a result, it is possible that in the coming years the entire organization chart will be altered to make way for new structural relations and positions.

Such a trend has been developing for several years, is accelerating due to the influence of the new technologies, and is a basic reason why network architectures find themselves at a crossroads. For one part, they must assure the compunications requirements of an organization in an efficient and timely manner, within an ever-expanding applications horizon. To datacomm was added text, then image and voice, and to complicate the picture even further, these have classically been areas of organizational responsibility under different executives with quite diverse (and often incompatible) goals.

Network architectures are also faced with their own problems of compatibility and internetworking, and they are subjected to many influences, not the least being that of technology. Satellite lines, for instance, bring the need to alter and adapt the Arpanet protocols which, based on the pioneering work of Baran (for packet switching) and Sutherland (in project management), do not reflect the requirements and capabilities satellite transmission has brought around.

When we talk of message handling and of the network itself as a transport mechanism, a primary interest is properly allocated to *flow control*, a basic problem for a link control procedure: how to match the sender's transmission rate with the receiver's ability to accept traffic. An explicit allocation of resources is a good solution.

The receiver clearly notifies the sender of its ability to accept messages, and the flow control routines aim at preventing the data sender from overflowing the receiver's buffers. Such a mechanism may exist at each level of protocol as well as between protocol levels; it assumes that flow control information is passed from the receiver to the sender and this information typically reflects the receiver's ability to buffer data.

To perform this task efficiently, the network architecture must have the means to represent a count of resources: unit of buffering, message queue elements, and so on. A good flow control scheme must handle a whole spectrum of problems that result from preventing buffer overflow in the receiver. The goals and methods of flow control include:

1. *End-to-end flow control:* Supervision for a particular level of protocol should be exerted at the point closest to the final destination.

2. *Congestion prevention:* The flow control strategy should reasonably guarantee that the traffic is not saturating the channels and that the contention for the available resources is kept within acceptable limits.

The control of congestion is one of the most important objectives in designing message- and packet-switching communications networks. Congestion occurs when the rate of arriving traffic exceeds the service rate provided by the network. Two types of control procedures can be implemented:

- *End to end*, which places restrictions directly at the message source by monitoring the number of messages on the logical connection: source to destination.

- *Local*

Neither end-to-end control nor local control, in isolation, is a complete solution to the congestion problem, but combined they help solve it.

Experiments have also demonstrated what alternative solutions can offer. Probabilistic routing seems to be better than deterministic routing, in terms of network congestion, for locally controlled networks. Including random routing may enhance its usefulness in providing an analysis indicating alternating routing performance, but the most critical consideration is the steady, proper collection of traffic statistics to provide a factual and documented basis for optimization.

Message switching brought forward, and packet switching adopted and extended, the concept of *store and forward.* This is a great development not only because it helps in solving congestion problems, but also because it made feasible the dissociation of sender and receiver. Gone is the requirement that sender and receiver must be online at the same time through a physical line capability. For text and data, but also for voice, we can store at any node the message as it arrives and forward it at will (or at a preestablished time).

Store and forward messages are typically terminal-to-terminal or unsolicited host output, stored on disk and routed to the destination. In a distributed network architecture, these messages involve a wider variety of destinations than does centrally switched traffic, but they also require more extensive routing capabilities.

For a centrally switched traffic, routing functions are necessary to extract the destination logical identifier, determine the destination line and station tables, complete the control block for output with the appropriate routing data, and queue the control block for output. Similar functions are necessary for node (distributed)-switched traffic, but the following functions must also be provided:

- *Edit:* Extract the routing data from the message content.
- *Route:* Use the routing data to determine the logical identifier and then the line and station tables.
- *Tag build:* Complete the control block and build additional control blocks as required for multiple deliveries.
- *Queue:* Line up the control blocks, send message acknowledgments, write logs as necessary, and dequeue.

These are primitives for store and forward. A typical queuing program writes the control blocks to the appropriate output queues, updates the queue pointers, sends acknowledgments when required, sets the flag for output service action, and performs logging. Similarly, an output queue service routine reads the control block entries from the queues, reads the message segments, and interfaces with the appropriate input/output routine to initiate output.

Understandably, all these operations are transparent to the user. They interest the network architect and the designer, but they also help identify the relative complexity of a network architecture. The routines supporting it must be updated steadily with the state of the art. This causes the successive "Releases." We will take IBM's SNA (System Network Architecture) as an example.

SYSTEM NETWORK ARCHITECTURE

IBM's development of a System Network Architecture has come in stages: very little resemblance exists between Version 4.2 and the original Version 1. Indeed, many rumors persist in the industry that IBM is preparing to break away from SNA, and that equipment capabilities to support a different type of network architecture are being built around small mainframe technologies.

The introduction of SNA in the early 1970s aimed at answering the objective of a broad sharing of resources, allowing central computer equipment use for multiple functions. However, Version (Release) 1 did not achieve this objective, as it centered on a strictly hierarchical structure. The largely star-type network intermingled the basic functions

- Data processing
- Data transport
- Databasing

the same way that star methods had done in the late 1960s and early 1970s. There was no multicomputer network associated with Version 1, and the first SNA implementation was very restricted.

The only real novelty was SDLC (Synchronous Data Link Control), but there was also the limitation that only SDLC terminals could be employed. Furthermore, IBM made a serious marketing miscalculation in assuming that users would immediately migrate to SDLC. It did not happen that way.

Version 2 improved on Version 1 by allowing SDLC and bisynchronous terminals to coexist on the same network (and front end), although not on the same line. Yet although Version 2 relaxed some of Version 1's constraints, it did not alter the basic concept: SNA still operated on a single host system.

Version 3 changed this by allowing a multihost system through ACF (Advanced Communications Function). This made it feasible for terminals to access programs and database elements running on more than one host computer. (IBM talks of regions and domains, each domain being controlled by an applications host.)

The strictly hierarchical system survived with Version 3, and some other constraints from the original Version 1 also filtered through, such as the rigid routing capability. Next to ACF the most interesting innovation (for IBM) that Version 3 provided was that CMC (Communications Management Configuration) permits all terminals to be defined as being in the domain of one or more CMC hosts.

CMC was a rudimentary start on network management, but it was left to Version 4.2 (implemented in 1979) to answer the more sophisticated requirements imposed by local balancing and capacity sharing. Between Versions 3 and 4.2, IBM (in late 1978) announced new hardware, the 8100 and the 4300 Series.

Was the previously available hardware also wanting? Mainframes are not noted communications engines; they are batch machines. NPDA (Network Problem Determination Application) included in Version 4.2 is IBM's first significant approach with an online diagnostic feature. Used with the proper modems, NPDA can isolate malfunctions

- In the terminal control
- In the modem interface
- In line errors

Another enhancement with Version 4.2 is alternate routing, which resides in the network nodes. Data paths can be predefined in the system so that a failure in the primary transmission paths would cause automatic switching.

Version 4.2 has included as well software improvements: terminals in an SNA network that can operate more freely than Version 3 would ever allow. IBM is sure to have an advanced version under development. Two items will soon require corrected and added facilities:

- First, the issue of *interconnecting* SNA with the X.25 public networks

The interfaces to the French Transpac and Canadian Datapac networks have not worked as expected.

These networks (together with IBM's SBS) are now in advanced phases of implementation. If there are no overriding technical problems, all should be in operation within the next couple of years. To the contrary, SNA's path control is entirely different from X.25 and incompatible with it.

It will probably be expensive for IBM to rip out the path control of existing SNA machines and replace it with an X.25 approach. This is one reason several major firms have decided to keep out of SNA, at least for the moment. The other reason is that even with Version 4.2, SNA still offers some fuzzy interfaces. The third is the cost of the implementation.

- Second, the eventual integration of text, image, and voice

IBM has a stake in this because of its 3750 and its 1750 PBXs, but the potential impact of this development is much greater, as it eventually integrates office automation, teleconferencing, and datacomm into one information system for offices and households.

The most likely strategy is to try, with the new release, to make SNA the de facto standard in a distributed data processing/data communications environment integrating it into its PBX offering. The capability to interconnect all types of information-handling systems into coherent, user-friendly communications networks is a very important goal during this decade.

With regard to user acceptance, a primary future requirement will be that the majority of users be able to get in and out of other networks and other product options, centering around an international datacomm protocol such as X.25. We consider the impact of international standards in the next chapter.

Chapter 17

STANDARDS FOR SESSION CONTROL

INTRODUCTION

Processes running on a computer need to communicate with other processes that reside at hosts and workstations distributed in a network. Such communications will typically involve transport activities external to the host and interest the functions supported by the network (as described in Chapter 15) and other activities proper to the host itself. Session control is the host function that interfaces with the transport capabilities supported by the network architecture.

Figure 17-1 supposes a mainframe in a typical bank on which run in a multiprogramming mode a number of routines: current accounts (DDA), savings, foreign operations, stock exchange, and general accounting. Through the datacomm engine attached to the host (front end, computer gateway) these activities will communicate with similar ones in the branch offices. The teller terminals work on a bisynchronous protocol (BSC); cash dispensers are asynchronous (start/stop); and the backoffice mini works on packet switching.

These protocols interest the lower layer of the communications (transport) facility. Supposing that the network follows the X.25 protocol, its interfaces (message processors) will be responsible for translating

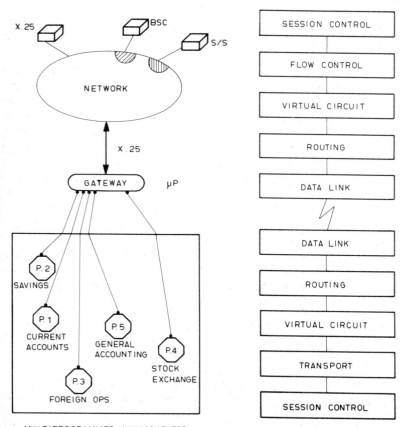

FIGURE 17-1 Processes run on a main-frame communicating through a datacomm engine and a network capability with terminal devices and other processes.

FIGURE 17-2 Layers in a communications network, from data link to session control.

the BSC and asynchronous protocols to HDLC (higher-level data link). For the system to work, the formalisms and conventions shown in Figure 17-2 must be observed.

Modern network architectures are typically layered. Each layer at a typical node includes the needed routines that follow the protocols and provide the desired network services at that node. In general, the functions are:

- *The physical link layer:* assures a data link with a physical transmission facility between adjacent nodes

LEVEL 10	BASIC SOFTWARE	FILE ACCESS	LOGICAL	
LEVEL 9	BASIC SOFTWARE	D B M	LOGICAL	
LEVEL 8	BASIC SOFTWARE	PROCESSES	LOGICAL	
LEVEL 7	BASIC SOFTWARE	PRESENTATION	LOGICAL	
LEVEL 6	BASIC SOFTWARE	SESSION	LOGICAL (ON TWO USERS)	
LEVEL 5		TRANSPORT	LOGICAL	NOT EXACTLY STANDARDIZED
LEVEL 4	NETWORKING	VC/D	LOGICAL	STANDARD X.25 (FOR VC)
LEVEL 3	NETWORKING	ROUTING	LOGICAL	STANDARD X.25
LEVEL 2		DATA LINK	LOGICAL	NON-STANDARDIZED BUT TRANSPARENT
LEVEL 1		D C E	PHYSICAL	STANDARDIZED

FIGURE 17-3 A distinction between physical and logical layers (levels) with indication of international standards, where applicable.

In Figure 17-3 this is indicated as *Layer 1* (Layer 0 is the data terminating equipment, DTE). This is the function of the modem (data communication equipment, DCE). Several modules may be specified for this layer (such as AT&T's RS232C and CCITT), one for each type of transmission facility.

- *The data link layer* (Level 2): guarantees an error-free communications mechanism between adjacent nodes

The way data flows in a network is determined by a significant characteristic of the two lower layers. The modules that reside in the physical link and data link layers provide services only for moving data from a given node to an adjacent node.

- *The transport layers* (Levels 3, 4, and 5): provide a routing mechanism for the network services, also virtual circuit and flow control. The supported functions allow a data packet to be sent from any node to any other node in the network.

The transport layers and the data link come under different names in different architectures, as Figure 17-4 documents. Neither are the sup-

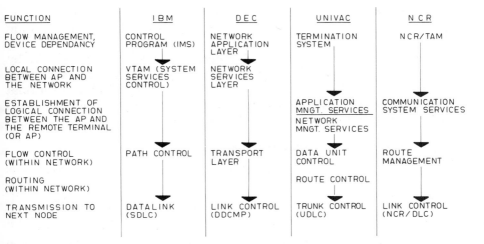

FUNCTION	IBM	DEC	UNIVAC	NCR
FLOW MANAGEMENT, DEVICE DEPENDANCY	CONTROL PROGRAM (IMS)	NETWORK APPLICATION LAYER	TERMINATION SYSTEM	NCR/TAM
LOCAL CONNECTION BETWEEN AP AND THE NETWORK	VTAM (SYSTEM SERVICES CONTROL)	NETWORK SERVICES LAYER		
ESTABLISHMENT OF LOGICAL CONNECTION BETWEEN THE AP AND THE REMOTE TERMINAL (OR AP)			APPLICATION MNGT. SERVICES NETWORK MNGT. SERVICES	COMMUNICATION SYSTEM SERVICES
FLOW CONTROL (WITHIN NETWORK)	PATH CONTROL	TRANSPORT LAYER	DATA UNIT CONTROL	ROUTE MANAGEMENT
ROUTING (WITHIN NETWORK)			ROUTE CONTROL	
TRANSMISSION TO NEXT NODE	DATALINK (SDLC)	LINK CONTROL (DDCMP)	TRUNK CONTROL (UDLC)	LINK CONTROL (NCR/DLC)

FIGURE 17-4   A comparison of network functions supported by different network architectures.

ported functions exactly the same, but basic principles are observed: a module may use the services of a module in a lower layer but not those of modules in the same or higher layers; two modules in the same layer but residing in different nodes would cooperate to fulfill their network functions by exchanging the appropriate protocol messages.

- *The session control layer:* guarantees a location-independent communication mechanism, taking care of the logical aspects of the transmission

Two application modules may communicate with each other by means of this layer, regardless of their locations in the network. We will examine the functions of session control in considerable detail.

- *The presentation control layer:* provides the finer programmatic interfaces for the user layer

Presentation control follows up on the establishment of a session, maintaining and terminating it, including the function of requesting process creation; it delimits data enclosures, prepares the necessary delimiters for addressing workstations via mailboxes—a function to be accomplished by session control; and notifies processes upon receipt of data and data enclosure delimiters.

In simple terms, presentation control makes data more understandable through segmenting, blocking, and so on; provides for message management beyond the level of buffering and controlling data (which is proper to session control); actuates the commitment process; provides credits/debits for guaranteed delivery; and supports data operations beyond checkpoint, recovery, and commitment (proper to session control).

The protocol followed in presentation control is a set of rules by which a session is assisted in establishing, maintaining, and terminating a data transfer. The interface to the session layer includes the format by which control information is passed and the rules to be interpreted to transfer data. Presentation control needs to:

- Make data more understandable.

- Assure a reasonable homogeneity of languages between processes and DTEs.

- Adapt requests to the specific machinery existing in that location if virtual terminals are used.

- Translate these names to a common reference if programs use local names.

Presentation control must further assure the necessary compacting/decompacting, provide for information enrichment, and guarantee encrypting/disencrypting when required.

THE OBJECT OF SESSION CONTROL

There is today a standard supported by the International Standards Organization (ISO), known as the Open System links connection (OSI). The ISO/OSI is communications-oriented but resident in the Data Terminating Equipment, not in the network. It is layered and normalizes two layers: *session* and *presentation* control.

The object of session control is to take care of the logical aspects of the transmission from one process to another, whether local or remote. This includes the routines in the applications library, those in the utilities library, and the database elements.

Rules and supports are necessary to control and interpret the operation of different devices at the workstation level, execute commands on the behalf of the processes, receive commands from the correspondent processes associated with the session, and pass them on to the process of the workstation. The set of rules by which a session is established, maintained, and terminated includes the logical aspects of the transmission.

Control information must assure log on/log off and a number of programmatic interfaces. The latter regard the rules by which a human-information dialogue is interpreted to make available to the user the data he or she asks for: send, receive, establish, maintain, and terminate are activities largely falling under this scheme.

Session control complements the networking operations (data paths, routing) and provides the section of transport control nearest the user. Among the basic functions we distinguish *user support* to dynamically redefine part or all of his or her display environment; the capability to *tailor the operation* of the terminal to the user's personal needs; and the provision of software to receive input from, and place its output in, *session control.*

Some technical definitions are necessary to assure consistent user interfaces: terminals, word processors, databases, information elements, processes, and administrative languages.

A stream can be thought of as a virtual sheet of paper upon which we can both write and read. The establishment of a session presupposed a *connection,* that is, a logical association capable of transferring data between DTEs or processes. *DTE* (data terminating equipment) is an addressable end point in a location: a place where things happen (principally, terminal, workstation, host).

A *data unit* is a quantity of data and control information transferred as a unit over a connection. A *data assurance unit* is a quantity of data whose successful transfer over a connection is acknowledged. A *data mapping unit* is an entity used to map a data unit of the next higher level onto a data unit of the current level.

A *process* supports application and system activities; this definitely includes exchanges of information by which cooperation is achieved, end to end, with other processes. A session *quarantine* unit is a quantity of text that cannot be released to the recipient process until the sending process signals its completion.

Session *interaction units* are a subdivision of a session commitment delimited by the passing of control of a session from one workstation process to its correspondent process. Whoever is in control of the session may request such services as termination or other types of interaction.

Some other ANSI-sponsored, standard definitions are: session *commitment* unit—a quantity of data transferred during a session; session *recovery* unit—a quantity of data transferred during a session—a *commitment* unit; session *data* unit—a unit of data transferred between a pair of processes using a session. For the manufacturer, the philosophy of the systems architecture which it is adopting and implementing constitutes the future reference for the evolution of its product line: from mainframes and terminals to operating systems and applications routines. The criticality of this issue cannot be dramatized better than by saying that for the next decades the architecture will condition the service to its clients.

At how many levels must a solution be examined?

- The first reference is that of the basic concept.
- The next concern is the technical definition of a set of rules.
- Then come the architecture-based products, as the successive releases of this architecture.

The objectives of looking at a concept are to project a map of what the system should and could do: to provide a base on which users can configure systems to meet requirements; to assure a transition path from current solutions to local networking; to evaluate the possibility of internetworking with the developing VAN (public value-added networks) for long-haul interconnection; and to make it feasible to implement a layered functionality involving:

- Network characteristics
- DTE
- Topological distribution

Technical support should see through an NCC capability, including all necessary administrative duties involved in running the network, and it should be possible to distribute the status and error information at the level of the component itself.

For internetworking reasons, it is advisable that the system architecture follow the ISO approach on "Open Systems" characteristics, including the definition of the Activity Boundaries of the DTE and of the hosts. This involves the support of:

- *Ports*, which are hardware bound.
- *Sockets* (or end points; entities, E) with both S/W and H/W boundaries. Sockets act similar to a telephone number: a software convention, conditioned by H/W characteristics.
- *Logical communication paths*, which connect socket to socket (Figure 17-5).

The S/W–H/W boundaries assure that the transfer of data will follow formatting rules. Entities (E) and virtual terminals (VT) are two layers within the broader terminal management (TM) layer. The goal of the virtual terminal service is to establish the formatting rules common to all communicating DTE on the network.

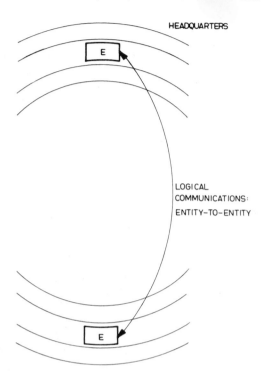

FIGURE 17-5   An end-to-end (entity-to-entity) logical communications path.

In different terms, a major reason for having TM is to provide a generally supported virtual terminal capability. Within the framework of the *logical terminal* the TM layers fit between session and presentation control on one side, and AP (applications programs) to DB on the other. The latter are higher-up layers of TM (Figure 17-6).

A number of routines are necessary to support the online communications characteristics of the system. These may be the common exchange interface to handle the communications characteristics between the flow control layer (which belongs to the networking side) and the session control layer, the point of entry to the terminal or workstation.

Typically, session control will define authentication and privacy for all exchanges. Next to it, presentation control will take care of the finer programmatic interfaces, among them encryption. A formatted form of data transfer can be assured through a standard device protocol interface. The way presentation control talks to the AP (applications programs, programmed products, procedures) is conditioned by these routines.

Figure 17-7 demonstrates this point. Say that different processes,

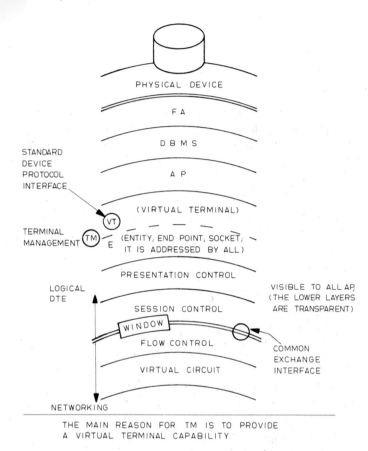

THE MAIN REASON FOR TM IS TO PROVIDE
A VIRTUAL TERMINAL CAPABILITY

FIGURE 17-6   Explosion of the terminal management capability into two layers: entity and virtual terminal.

some running on a mainframe, others at a dedicated minicomputer level, concern three procedures:

- Personnel (AP 1)
- General accounting (AP 2)
- Billing (AP 3)

AP 1 at a minicomputer needs to communicate with AP 1 at the host. A *logical* connection will be established to assure this communication. Everything else (in terms of subordinate layers) is transparent to this logical connection. The same is true of the other processes and their interconnection.

The VT layer helps define how terminal management looks to the

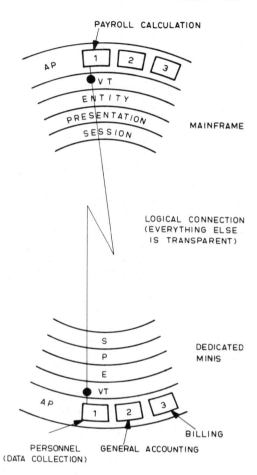

FIGURE 17-7 Applications processes (AP) in a mainframe communicating with similar applications processes in a dedicated mini, through session and presentation control, entity, and virtual terminal.

inside of the system. This may include several references: editing (CRT), graphics, and so on. Toward the higher-up layers, the TM action can be based on a given schema of the database. In facing the lower-level layers (communications), TM includes the drivers that worry about the exact specifications of each terminal and workstation type.

OPEN SYSTEMS INTERCONNECTION

Open Systems Interconnection (OSI) refers to procedures for exchanging information among terminals, workstations, computers, processes, people, networks, and so on, that are *open* to one another for this purpose by virtue of their mutual use of these procedures. "Being open" refers to the mutual recognition and support of standard information exchange procedures.

In this concept, a system is a set of one or more entities and associated software and hardware that form an autonomous whole capable of performing information processing. OSI standards are concerned with exchanging information between systems, not with the internal functioning of each individual system.

The transfer of information between systems is performed by physical and logical media for systems interconnection. The media initially considered in developing OSI standards are of the telecommunications type. Furthermore, OSI is not only concerned with transferring information between systems in pure communications terms, but also with their capability to interwork to achieve a common, distributed task.

Applications processes are basic OSI elements and illustrated by such entities as logical or physical processes. Architectural principles may be applied: the concept of a layered architecture, with layers, entities, service access points, protocols, and so on. Identifiers must be introduced for entities, service access points, and connections; protocols must be considered and with them general architectural elements, including connections, transmission, and error functions. Management aspects, routing issues, and security constitute legitimate parts of the architecture.

The basic structuring technique in the OSI architecture is layering: each system is viewed as being logically composed of an ordered set of subsystems, represented for convenience in a vertical sequence. Adjacent subsystems communicate through their common interface. Subsystems of the same rank collectively form the X-layer of the architecture. Entities exist at each layer; those in the same layer are termed *peer entities*.

It may be necessary to further divide a layer into small substructures and to extend the technique of layering to cover other OSI dimensions. This is for further study, with possible techniques as sublayering, protocol switching, and the use of a multidimensional representation. A sublayer is a functional grouping in a layer; it can be bypassed in the establishment phase as well as in the data phase.

The applications layer is the highest layer in the reference model of OSI architecture. Protocols of this layer directly serve the end user by providing the distributed information service appropriate to an application to its management and to system management. OSI management comprises those functions required to initiate, maintain, terminate, and record data concerning the establishment of connections for data transfer among applications processes. The other layers exist only to support this layer.

An application is composed of cooperating *applications processes* which intercommunicate according to application layer protocols. Applications processes are the ultimate source and sink for data exchanged. A portion of an applications process is manifested in the applications layer as the execution of applications protocol (applications entity). Applications or applications processes may be of any kind (manual, computerized, logical, or physical).

The presentation layer provides the set of services which may be selected by the applications layer to enable it to interpret the meaning of the data exchanged. These services are for managing the entry, exchange, display, and control of structured data. The presentation service is location independent and is considered to be on top of the session layer, which provides the service of linking a pair of presentation entities.

The session layer assists in supporting the interactions between cooperating presentation entities. To do so, the session layer provides services such as binding two presentation entities into a relationship and unbinding them, a session administration service. Another example is the control of data exchange, delimiting and synchronizing data operations between two presentation entities, a session dialogue service.

To implement the transfer of data between presentation entities, the session layer may employ the services provided by the transport layer. The latter exists to provide a universal transport service in association with the underlying services provided by lower layers. The transport service provides transparent transfer of data between session entities; it relieves them from any concern with the detailed way in which reliable and cost-effective transfer of data is achieved.

The transport service is required to optimize the use of the available communications services to provide the performance required for each connection between session entities at a minimum cost. The lower network layer assures functional and procedural means to exchange network service data units between two transport entities over a network connection. It provides transport entities with independence from routing and switching considerations, including the case where a tandem subnetwork connection is used.

The data link layer addresses itself to the functional and procedural means needed to establish, maintain, and release one or more data links among network entities. The physical layer provides mechanical, electrical, functional, and procedural characteristics to establish, maintain, and release physical connections between link entities.

Each service provided by an X-layer may be tailored by choosing one or more facilities which determine the attributes of the service. When a single entity of a layer cannot by itself fully support a service requested by one of its next-higher layer entities, it calls on other entities in the same layer to help complete the service request. To cooperate, entities in any layer other than the lowest communicate through the set of services provided by the next lower layer. The entities in the lowest layer are assumed to communicate directly via the physical media that connect them.

An entity may provide services to one or more entities in the next-higher layer and use the services of one or more entities in the next-lower layer. A service access point is the access means by which a pair of entities in adjacent layers use or provide services. The cooperation between entities in the same layer is governed by a set of protocols specific to the layer.

A *global title* identifies an entity regardless of its location and is unchanged if the entity moves in any manner. A *local title* is a name which uniquely identifies distinct entities only within a limited context. The resolution of a local title to the global title of an entity is a function defined within the architecture. The context of a local title is its title domain.

A global title consists of two parts: a *title domain name*, which identifies its title domain uniquely in the OSI environment, and a *title suffix*, which is unique within the scope of the title domain so identified. Global titles have been defined as structured names to facilitate their administration. They permit the orderly delegation of authority to assign new global titles and to create new title domains which are subsets of existing ones. The layers themselves are the most important title domains.

Using an address to identify an entity is an efficient mechanism if the permanence of attachment between the entity and the service access point can be assured. If there is a requirement to identify an entity regardless of its current location, the global title assures the correct identification.

Two different kinds of mapping functions may, in particular, exist within a layer: hierarchical and table. A hierarchical structure of addresses within a given layer simplifies address mapping functions within that layer because of the permanence of the mapping it supposes. Mapping by tables is a procedure which has been used since the early days of computing (function, arguments) in computation and file access. Its employment in networking and databasing is a more sophisticated use of the same principle.

A *routing* concept is defined within the architecture. It serves as a function to translate the address of an entity into a path or route by which the entity may be reached. Control information and user data are exchanged between entities in protocol data units which contain protocol control information and possibly user data.

Connection multiplexing is a function with the layer by which its connections are mapped. The mapping may be one to one, many connections to one connection (upward multiplexing), or one connection to many connections (downward multiplexing). Upward multiplexing may be needed in order to:

- Make more efficient or more economic use of the service.
- Provide several connections in an environment where only a single connection exists.

Downward multiplexing may be needed for reliability because more than one connection is available, to provide the required grade of performance by using multiple connections, or to obtain some cost benefits by using multiple low-cost connections, each with less than the required grade of performance. These connection mapping functions involve a number of associated functions that might not be needed when one-to-one connection mapping is done.

Upward multiplexing comes in perspective to ensure that user data from the various multiplexed connections are not mixed. When the capacity of the connection is shared by the introduction of a multiplexing function, it is necessary that flow control functions be performed on each individual flow.

Two types of flow control are identified: peer-to-peer flow control, which requires protocol definitions and is based on protocol data unit size; interface flow control between adjacent layers, which is based on the interface data unit size. Multiplexing in a layer may require a peer-to-peer flow control mechanism for individual flows.

Segmenting and blocking may be needed, as data units in the various layers will not necessarily be of compatible size. Segmenting may occur when protocol data units are mapped into interface data units. If blocking is performed, several service data units with added protocol control information form a protocol data unit.

An acknowledgment function may be used between peer entities using a protocol to obtain a higher probability of loss detection than provided by the layer. Each protocol data unit exchanged by corresponding entities must be uniquely identifiable so that the receiver can inform the sender that the protocol data unit was received. The acknowledgment functions must also be able to infer the nonreceipt of protocol data units and take appropriate remedial action.

The scheme for uniquely identifying X-protocol data units may also be used to support other functions, such as duplicate detection, segmenting, and sequencing. Protocol may use certain *error control* functions to provide a higher probability of protocol data unit errors and data corruption detection than is provided by the layer. Error control and notifications may require that special information be included in the data unit's protocol control information.

Some services will require a *reset* function to recover from a loss of synchronization between correspondent entities to come back to a predefined state with a possible loss or duplication of data, while additional functions may be required to determine at what point reliable data transfer was interrupted.

Systems management is considered a special applications processing case consisting of system management entities distributed over the interconnected Open Systems, which communicate with each other to coordinate their activities. Their communications must follow, as much as possible, the general rules defined by the OSI architecture for all the applications. Connections between management entities should be established when a system that has been operating in isolation from other systems becomes part of the Open Systems Interconnection architecture.

SESSION CONNECTION SERVICES

As discussed in a previous section, the object of Session Control is to enable two entities to establish a connection between themselves. This includes the means to request a connection to another presentation entity; receive a request for session connection from another presentation entity and accept, negotiate, or reject it; and be notified of connection establishment or the rejection of the session request.

Parameter management by session control allows the presentation entities to determine the connection's unique value. Before starting the transmission between themselves, the presentation entities must agree on a set of parameter values concerning the session layer (session characteristics) and the presentation/applications layers. Typical parameters may be associated with local session end point identifier, interaction management, control data exchange, session synchronization, expedited data, quarantining, and user data exchange. The agreement on session parameters may encompass:

- The definition or negotiation of an envelope that determines the possible range of values acceptable by the presentation entities

- The selection, in this envelope, of the particular value of the parameter to be used

Such actions are generally performed at session connection establishment; the latter may be used to reselect parameters during the data transfer phase. In negotiation, the session layer determines values mutually agreeable to the two user correspondents. In imposition, either the values proposed by the initiating entity are accepted or the session/ connection request is rejected.

The session identification service provides to the presentation entities (1) a local session end point ID to refer to the connection during its lifetime, and (2) an identifier that uniquely specifies the connection within the environment of the cooperating entities, with a scope greater than the lifetime of the connection. This identifier may be used for administrative purposes: accounting, diagnostics, and recovery.

The session release service allows the presentation entities to release a connection. The release can be orderly, that is, cooperative with no data lost; or destructive (aborted). In this instance, one presentation entity unilaterally releases the session; this may cause loss of data. Connection service provides for request release of the session/connection (orderly or destructive) and receives an indication of connection release.

The session service data unit is the smallest unit of data whose enclosure is preserved from end to end between the presentation entities. This enclosure is used for data structuring and made known to the receiver but does not affect the functions within the layer. Such service includes the assurance that the receiving presentation entity is not overloaded with data. A data quarantine service allows the sending presentation entity to request that a series of data units sent on a session connection not be delivered to the receiving entity until explicitly released by the sending entity. The sending presentation entity may request that all the data currently quarantined be discarded.

Quarantine service parameters are determined during session/connection establishment. Quarantining may be accomplished by establishing whether quarantining is "on" or "off" as a session parameter. Quarantining service parameters may be changed during data exchange by these types of send/receive interaction involving session service data units:

- *Two-way simultaneous interaction:* Both presentation entities may currently send and receive.

- *Two-way alternate interaction:* The presentation entity with the turn may send; its correspondent is permitted only to receive.

- *One-way interaction:* Only one presentation entity may send; its correspondent only receives.

This service provides for determining initial ownership of the turn and the type of interaction and also assures voluntary or involuntary exchange of the turn. In the case of involuntary exchange of the turn, data may be lost. Turn assignment does not influence the exchange of expedited data units, nor is it necessary to transmit acknowledgments of turn exchange.

Initial ownership of the turn may be negotiated as part of session/connection establishment. Alternatively, initial ownership of the turn may be assigned to the presentation entity that initiates the session connection. A session synchronization service assists the presentation

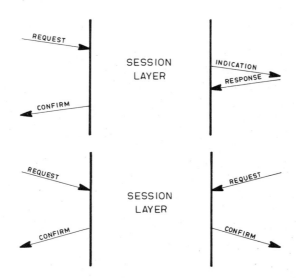

FIGURE 17-8 Handshakes for establish, synchronize, and release operations relative to the session layer.

entities in synchronizing their dialogue; it provides for marking and acknowledging of identifiable synchronization points and for resetting the session to a defined state and agreeing on a defined resynchronization point.

Parameters associated with service primitives are independent of the characteristics of a particular interface. A *primitive* is a fundamental operation (native command) supported by the software. Event diagrams indicate the logical relationship of such events constituting a service primitive. There is a taxonomy of events, the four types being: request, indication, response, and confirm.

A *request* is an event that initiates a service. It is unprovoked and sets in motion the series of events associated with a service primitive. An *indication* is an event usually provoked by a request which notifies a user that a service has been requested. It may or may not call for a response. An indication appears independent of previous events at a single interface.

A *response* is an event initiated by a user in answer to or in cooperation with an indication-type event. It differs from a request in the relationship it bears to the indication. *Confirm* is an event that indicates the completion of a service. The completion may be successful or unsuccessful, and this is normally indicated by the *result* parameter associated with confirm-type events. Confirm events expect no related response-type events.

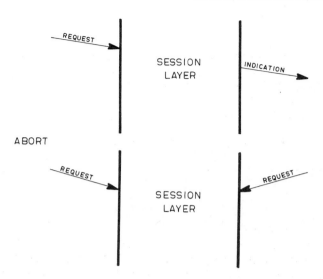

FIGURE 17-9   Handshakes for abort, quarantine, and turn management relative to the session layer.

Establish, synchronize, and *release* operations are presented diagrammatically in Figure 17-8. In the establish case, connect collision should result in successful establishment of a single session connection. If the response to a synchronize operation is *reject,* it may be accompanied by a counteroffer in the form of another synchronize request.

An *abort* operation is exemplified in Figure 17-9. The first half of this figure is also valid for *quarantine* and *turn management* operations, which constitute the primitives of the session service data unit function. An expedited data exchange service provides the means by which a sending presentation entity identifies one or more data units for transmission to the receiving presentation entity more rapidly than data transmitted via the user data exchange service. Such expedited data may overtake normal data. Although the amount of normal data overtaken cannot be predicted or guaranteed, it is guaranteed that expedited data will not be delivered to the receiving presentation entity later than data sent subsequently. Specific restrictions are placed on the size and number of data units transmitted at one time by the expedited data exchange service.

Formalisms characterize all network architectures. They may not be identical to the international standard we have been describing, but they do exist in some form or another to guide the actions of the communicating processes, files, devices, or human beings who interact with one another through the network.

Chapter 18

LOCAL NETWORKS

INTRODUCTION

The new voice and data communications environment calls attention to the issue of local networks. There is a difference in primitives between long-haul (wide-area) and local networking, and the same is true of the distinction that exists between local networks and the internal computer structure (Figure 18-1). Long-haul, local area, and internal structure are, in fact, three basic references in network capabilities, each with its own applications field and technical requirements.

A different way of presenting this three-way classification is to bring under perspective remote computer networking capabilities such as Aloha, Arpanet, Telenet, Datapac, Transpac, and SNA; local networking, with Ethernet, Omninet, Cluster/One, ARC, Wangnet, Aran, and Localnet as examples; and multiprocessing, with Illiac and so many other maxicomputers.

Compunications grew from RT (real time), TS (time sharing), data communications, and multiprocessing requirements. The last led to networking, as networks allow engagement in distributed multiprocessing. As experience accumulated it became evident that substantial remodel-

THE NETWORK CONCEPT MUST
BE SEEN AT THREE LEVELS

| 1. LONG HAUL |

SAY, 100 TO OVER 5.000 KM, CHARACTERIZED BY
   A. MULTIPLICITY OF DATACOMM MEDIA
   B. SYNCHRONIZATION
   C. REPEATER PROBLEMS
   D. LACK OF DIRECT CONTROL

| 2. LOCAL AREA |

TYPICALLY: FACTORY
           OFFICE BUILDING
           APARTMENT BUILDING
SAY, 500 M TO 7 KM AND BEYOND

| 3. INTERNAL COMPUTER STRUCTURE |

MULTI ACCESS
MULTI PROCESSING
MULTI COMPUTING

CONTENTION PROBLEMS NEED TO BE
STUDIED IN EACH CASE

FIGURE 18-1   A three-way classification of an application of the network concept: long haul, local area, and the internal computer structure.

ing was necessary to achieve reliability, availability, modularity, good throughput, low delay, and good-quality service.

A local area network (LAN) architecture can be of many types; but whichever the type, it has to satisfy basic technical prerequisites. Their characteristics have been studied by standardization committees, such as the 802 Project of AIEEE, ECMA (the European Computer Manufacturers' Association), and ISO (the International Standards Organization).

Still, the degrees of freedom in developing and announcing a LAN architecture are actually quite large. Judging from the offerings which during the last three years have entered the marketplace, they range from low-cost networks for personal computers, to fast, high-volume links for mainframes. Some networks are proprietary and interconnect products from only one vendor, though most offerings are flexible and open to attached computer resources from different vendors.

Corvus Systems has introduced *Omninet*; *Cluster/One* Model A by Nestar Systems; Net/One comes from Ungermann-Bass Inc.; Z-Net from Zilog; a network design from Motorola; another one by Intel. The latter is a six-layer local area network in which the first two layers equal the entire Ethernet scheme.

The key area of application for local area networks is the office. The engine is the personal computer. It is a market imperative for personal computers and office automation vendors, since the ability of workstations to communicate will be the most important feature of future office systems.

THE AIEEE 802 STANDARD

Local networks can provide economical solutions in hardware, software, and installation costs for connecting existing computers and terminals, and if properly planned, they can alleviate communications bottlenecks. Some of the technical issues involved concern questions of how to address the machines in the network, and how the text and data packets travel. These issues have solutions and since the materials cost of the cable is so low, when offices are being built or refurbished it is worthwhile to install local network cable which can be tapped later when device requirements and locations are known.

Processors hooked on a local network can exchange text and data at speeds as high or better than those achieved between tasks or processes in most multitasking or time-sharing systems. Local networks do not suffer from as high a level of swapping and scheduling as that common in central computer–based operations. Hence text and data can be distributed as the requirements imply and the architecture permits leveling loads and improving performance among processors attached to the network.

Is there a clear definition of a local area network? The answer is "yes!" According to the AIEEE Committee 802, "A LAN is a datacomm system allowing a number of independent devices to communicate directly with each other, within a moderately sized geographic area over a physical communications channel of moderate data rates."

The mission involved in this 802 Project by AIEEE has been divided into three subcommittees:

1. Media
2. Access Control
3. Higher-Level Interface

while smaller groups addressed themselves to:

4. Functional Requirements
5. Glossary
6. Draft Organization

The functional requirements for the transport mechanism were designed to support media and topology independence, a functional independence to be implemented as low as possible in the architecture, direct party to party communication, coexistence and interoperability, OSI standards cognizance, and fairness criteria. Most importantly, the projected applicability of these standards was to involve not only the classical data processing but also file transfer, DB access, graphics, word processing, teletex, and digitized voice.

Typical devices under consideration for LAN attachment are:

- Computers of all types and sizes (maxi, mini, micro, nano)
- Terminals
- Plotters/printers, and so on

The scope is typically baseband: in the 1 to 20 Mbps range; accepting over 200 connected devices; at a lesser than or equal to 2 km distance; featuring code independence; single/multiple delivery; and simultaneous servers. Let's add that between LAN and long-haul networks (LHN) can be located an intermediate class: the "Metropolitan Area Network" which typically covers a 25 to 35 km range.

With LAN, the 802 AIEEE Committee admitted that there are really two distinct areas of operation:

1. Light load, low delay, low throughput
2. Heavy load, long delay—with throughput going from higher to lower through a controlled degradation

The basic concept of any transport mechanism is that, because of concurrency and parallelism, as we add more stations on the LAN the network becomes congested. Therefore, it is not at all true that a network working at double the speed can handle twice as many WS in the environment. As a result, we must be very careful in handling multiple services such as voice and data.

One service, for instance voice, may monopolize the network and bring up imbalance. To start with, the transport requirements are different: voice traffic can last 2 to 3 minutes—while data packets last for a few seconds. Hence, voice monopolizes the network, shuts the data pipeline, and upsets the services the network should offer.

Typically, local area networks operate within a site or building, providing high-speed (megabits per second) and low-error-rate communications between computers at relatively low cost. Speeds lie between those of long-haul networks, using dial-up or leased telco circuits operating at speeds up to a few thousand bps, and the speeds of internal computer buses, clocked at about 100 megabits per second.

Monitoring is crucial. Good network architectures are built to be monitored: they must keep track of every packet going by. Networks monitor the traffic through gateways, but at the current state-of-the-art it is up to the user to write the software which can store quality histories, and be able to "move back in time" and examine certain events. In short, the challenges are many and the industry's experience is not yet very detailed.

Research has steadily emphasized the need for a control theory approach to handling packets between local nets and within a local net. In an environment where word processors, data processors, and voice stations may coexist, a basic need is for buffers, as diverse devices working at different speeds and calling for efficient interfaces.

We want the text and data to be at the user's end, not stuck somewhere in the network because of bottlenecks and incompatibilities. Therefore, a careful study should consider four key topics:

1. The mix of hardware, software, and applications running on the network

2. Performance data projected by the analysis and assured by management

3. Protocols to be chosen

4. The needed bandwidth

Companies now working on local networks have demonstrated that developmental work should serve the purpose of a prototype for the future office information systems, be open to distributed information systems (DIS) and electronic mail applications, assure the capability to access not only a central but also special databases, be served through a software distribution facility, and guarantee monitoring and control by the network control center.

AN 802 REFERENCE MODEL

We referred earlier to architectural considerations. Several of the more interesting local area networks have fully distributed packet-switching control, though there is as yet no single approach for a local network architecture dominating the others. As stated, several standard groups, such as

- The Institute of Electrical and Electronic Engineers (IEEE)
- The National Bureau of Standards (NBS)
- The American National Standards Institute (ANSI)
- The International Federation of Information Processing Societies (IFIPS)
- The European Computer Manufacturers Association (ECMA)
- The International Standards Organization (ISO)

are working on this problem. However, most local networks use packet-switching, and since there exists an international standard for network communications, X.25, most vendors make certain that their software observes X.25 rules.

NBS is working actively on local networks and has recently adopted new standards that deal with bit-oriented synchronous data transfer among computers and other buffered attached data communications devices. IFIPS has established a working group to study local networks, and an ANSI committee has been set up as a task group to develop proposals on standards for local networks. By means of this X3 Technical Committee, ANSI is trying to standardize interfaces for networks functioning at high speeds, above 50 Mbps, over distances greater than 1 km.

To my judgment, the most advanced as a standard is the 802 LAN Reference Model composed of physical and logical layers. The physical layer, which in OSI represents the DCE (data communication equipment, modem), is split into two sublayers:

- The lower one addresses itself to the *cable*. The cable is the medium.

The cable may be of many kinds: twisted wire, flat wire, coaxial, optical fibers.

- The upper is the media access unit (MAU).

The object of the MAU is signaling, encoding, and medium handling.

At the first logical layer level, the ISO/OSI need for data-link control (DLC) is answered through HDLC. In Project 802 this has been substituted by two sublayers:

- The lower one concerns media access control (MAC).

MAC interfaces between the logical link and the media access unit, thus assuring continuity between the physical and the logical functions:

- The upper layer is the logical link control (LLC).

The goal of this sublayer is the effective realization of the data transfer.

At the next logical level with ISO/OSI, the networking layer is divided into two sublayers: routing (the lower one) and virtual circuit (the upper). With Project 802, virtual circuit is substituted by the datagram. ISO/OSI, X.25, and the Project 802 employ packet-switching principles. There is the possibility of adding another sublayer at Level 3, that of "enhancement."

Finally, the transport layer, which is the highest in the datacomm side, has not yet been studied by ISO/OSI. It is not within the current mission of Project 802. ECMA 72 Class 4 addresses itself to it, but there are no well-defined results to be discussed yet. (Reference should also be made to two other ECMA committees: TC 24 deals with communications protocols, and TC 29 handles text interchange between WS.)

Media access, to which we made reference, is a subject where standardization is vital. In this area, the chosen AIEEE standards are pretty well defined to avoid wide swings away from guidelines.

1. *CSMA* (carrier sensing, multiple access)

For baseband coaxial cable CSMA works with collision detection (CD) capability. Broadband does not permit CD. CSMA/CD baseband can work up to 10 Mbps. (We will talk of broadband and baseband in the next section.)

2. *Token Ring*

The token ring can work baseband on twisted pair, or baseband with coaxial. The latter seems to have a practical limit at 4 Mbps. Slotted ring is not one of the standards whether in ECMA, ANSI, or AIEEE.

3. *Token Bus*

The token bus can work baseband on coaxial, or take a different approach: FSK (frequency shift key). There can also be a broadband coaxial solution.

BASEBAND AND BROADBAND

Coaxial cable is a common medium in computers and communications. CATV uses it extensively and CATV hardware carries many advantages: it is low costing, readily available, and very reliable, presenting itself as a solution to the problem of getting data, text, and image from one place to another.

Coaxial cable can work broadband (up to 400 Mbps) under current technology. Examples are Wangnet, Alan, and Localnet (Sytek). It can also work baseband (up to 10 Mbps) as the case of Ethernet and ARC demonstrate. Other baseband media are twisted wire and flat wire (ribbon cable).

Broadband and baseband are network transmission techniques. With *baseband* transmission, information is directly encoded and impressed upon the twisted pair or coaxial medium. Only one transmission signal at a time can be present on the medium without disruption.

Various media access control techniques exist, and we have reviewed those proposed as a standard in the preceding section. Baseband has the advantage of being simple to implement and maintain. Costs are low, and no FCC approval is needed.

Broadband transmission permits one or more signals to be present on the medium at the same time. Using radio frequency technology (FDM), the 400 Mbps bandwidth of a coaxial cable can be divided into individual channels, each allocated to a specific task.

Broadband local area networks can support thousands of users over many kilometers. Baseband networks can cover distances of fractions of a km and support up to 60 users (though a third of that is a more reasonable goal). As previously stated, several baseband LAN can piggyback on a broadband.

Using radio frequency techniques, broadband systems handle audio and video as well as text/data transmission. Broadband however implies a *headend* or central retransmission facility, needed for amplification and frequency translation. Also, interfaces to the network use expensive fixed-frequency agile (tunable) modems.

Costs are a consideration with broadband, but not a dramatic one since both headends and modems come off the shelf, the latter in the $4000 range. (Some analysts have noted that by using time division multiplexing at each node, broadband ports may be able to approach the $400 to $1000 cost, but FDM is the way that benefits from CATV experience.)

A $400 reference per attachment is important because it characterizes the baseband connection. On the other hand, as is so often noted, there is no reason why the two approaches cannot be used together:

- A baseband network within an office
- A broadband network tying the baseband networks together.

Open-eyed users are starting to realize the possibility, and manufacturers design the gateways to connect the two. Ungermann-Bass has announced a broadband version of its baseband Net/One local network and the ability to go from one to the other; 3M/Interactive Systems offers the capability to attach on its broadband Alan, the Omninet and Cluster/One—plus the bridge through which these two different networks can communicate.

Let's take notice that over local, metropolitan, and wider areas, text and data transfer may also be accomplished by other media, such as microwaves, laser beams, fiber optics, and infrared. Microwave links have been used for some time, but they may be too expensive for transmitting data only half a mile, and they are subject to government regulation.

Alternatives to the coaxial cable are actively sought and companies are coming forward with products designed to carry out short-range point-to-point communications outside the regulatory environment. Among these companies are:

- Datapoint (Light Link)
- IBM (infrared)
- Network Systems (Link Adapter)

Datapoint's "Light Link" is a full-duplex digital transmission using modulated infrared light. It has a range of 3.2 km in clear weather and 1.6 km under ordinary conditions, and it supports a data rate of up to 2.5 Mbps. Light Link operates as an ARC interprocessor bus, permitting applications and file processors in separate buildings to communicate data at full ARC interprocessor data rates.

IBM scientists (at the corporation's Zurich Research Laboratory) have demonstrated a novel way of transmitting computer data without wires in an enclosed space by broadcasting on infrared wavelengths. Such experiments are conducted with a 64-kbps infrared transmitter and receiver using modulation. The system employs a Carrier Sense Multiple Access with Collision Detection (CSMA/CD) contention capability.

In both these examples and in many others, infrared light communications can be used in an installation confined to one large room, or, alternatively, wherever a line-of-sight exists, without wires or cabling at all:

- An infrared signal emanates from a transmitter in the ceiling of the room
- Each terminal or device in the room which has a receiver is addressed by a separate code impressed on the signal emanating from the ceiling
- Infrared diodes in each device in the room respond with formatted signals that are detected by a receiver in the ceiling

Infrared systems can utilize line-of-sight links between nearby buildings, but these links are sometimes affected by rain, fog, snow, or pollution. On the positive side, such equipment replaces telephone lines and modems across distances of up to 1.6 km and at data speeds of up to 100 kbps (baseband)—though forthcoming products will allow data rates of up to 1.5 Mbps.

It is possible using fiber optics techniques to transmit data at rates in excess of 1 Gbps (1 billion bits per second). Currently available fiber optic cables offer bandwidths up to 3.3 GHz (1 billion cycles per second), and it is practically impossible to overload fiber optic cable. The problem arises because the connectors are costly and difficult to install.

An optical *tee* for making and breaking optical connections along a network path has not yet been developed without frequent *optical repeaters* to receive, amplify, and retransmit the optical signal; this increases installation complexity.

CHOOSING A NETWORK STRUCTURE

Contention control mechanisms have been worked out as a result of experiments with the Aloha packet radio system in Hawaii (sponsored by the Department of Defense). Aloha is not a local network; it is a shared radio network, but it can be used as a reference because it established the principle that if several transmitters have low duty cycles, it is efficient to have them simply transmit packets when they need to, taking a chance that there may be a collision between two or more packets transmitted nearly simultaneously.

The possible strategies for dealing with collisions are retransmitting a packet after a fixed or random time interval if an acknowledgment has not been received, sensing the traffic before expediting to find out if another packet is on the channel, and listening not only before but also while transmitting. The last approach is the CSMA/DC used in Ethernet and Cluster/One, among other LAN.

In general, the terminals or computers are connected to the local network through the Bus Interface Unit (BIU). This manages the packet's routing to the end stations, including circuit connection and error detection. Most local network architectures see to it that a single BIU can go down without affecting the rest of the system.

In a network with central control, the BIU is simply a conventional or limited-distance modem; the necessary logic for access to the network is in the terminals and the central computer. In a distributed information system (with ring control or a bus network with contention control) the logic in reference is primarily in the BIU. Hence its design has become a critical problem in distributed networks (Figure 18-2).

Contention control and bus interface are two examples helping to document that a local network has a control mechanism through which the connected terminals, word processors, and computers share the use of the bandwidth. A simple control method is multiplexing: the transmission of a number of messages simultaneously over a single circuit, which we have already discussed.

A network in which there is a separate cable pair (or other medium) per connected device is a space-division multiplexing application suitable for the kinds of local networks in which the rate per cable pair is limited to about 9.6 kbps. Newer systems with higher bandwidths employ both frequency-division multiplexing and time-division multiplex-

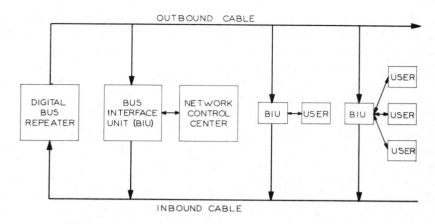

FIGURE 18-2   Functions of a bus interface unit in a local network.

ing. Whichever technology is used, the local network can be centralized or distributed.

A star network with space division of the communications channels, and certain networks with FDM or TDM (and fixed channel assignments to devices), are examples of centralized control. Distributed control can be developed for sharing bandwidth in wideband local networks. Ring and contention control call for the appropriate architectural considerations.

Let's recapitulate: The star arrangement, where there is one central device that provides the switching between all other devices, is the oldest approach. This is the configuration used in telephone exchanges and in many mainframe systems. In certain cases, such solution can be technically efficient, but if the dominant communications pattern is many sources to many destinations, the routing device in the star is quite likely to suffer from capacity problems. This is why when we spoke of optical fibers versus coaxial we suggested as an advantage of the latter the existing know-how for multitap on a ring-type network.

Star, ring, fully connected, and contention solutions are shown in Figure 18-3. Such network solutions will provide the backbone in the middle to late 1980s. The one to be chosen should support not only current requirements but also (and most particularly) those projected. These will typically involve:

- A growing number of interconnected, micro- and nanoprocessor-based workstations
- Video-oriented word processors
- An increasing integration of data processing terminals, minis, and small mainframes

- Image processing from management reporting to computer-aided design and manufacturing (CAD/CAM)
- Audio processing, integrating voice and, generally, sound systems
- Last, but not least, human factors, to be studied through human engineering

The factors behind the choice of a network structure are therefore complex and compounded by technical issues such as baseband versus broadband.

In general, user requirements, particularly in office information systems, are well served by distributed approaches. Demands for highly interactive solutions with very rapid response, and an increasing use of graphics and high-resolution text, can best be satisfied by providing personal workstations efficiently supported by the architecture. Two measures of efficiency are access to databases and the ability to communicate easily with other users.

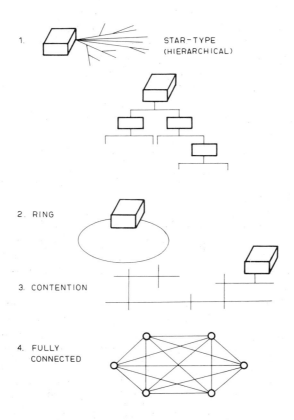

FIGURE 18-3   Four feasible local network architectures.

Another interesting technical issue is the recently developing claim by some vendors that their network architectures are IBM *SNA-compatible*. This claim is often misunderstood because there are two erroneous implications: first, that being SNA-compatible means that IBM's own network disciplines are less than proprietary, and second, that those competitor network architectures which do not present SNA compatibility are missing out.

At the current state of the art, no one is SNA-compatible. The claims are misleading and should read *SDLC-compatible* at the data link protocol layer, not at the higher layers of SNA. IBM is interested in promoting SDLC as *the* standard, and then anyone will be able to talk to an IBM system. But this talk is at a low level and, besides, there are many SDLC dialects on the market, not to mention HDLC, UDLC, BDLC, ADCCP, and so on.

The problem lies squarely with the local network's architecture ability to provide applications and system functions in a complex communications and database environment. The real compatibility challenge in the communications domain is not to IBM but to the telco value-added network: the local networks will act as subnetworks to that VAN (Figure 18-4).

Many issues must be taken into account in making a choice, not the least of which is tariffs. To optimize its system and provide for cost

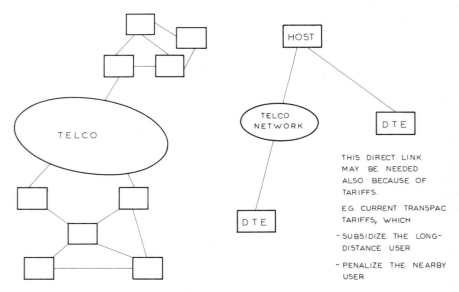

FIGURE 18-4   Subnetworking capability connected through a telco supported system.

FIGURE 18-5   A hybrid solution using both the telco network and private lines.

effectiveness, a company may be willing to adopt hybrid solutions (Figure 18-5). Furthermore, we cannot forget that the technical issues concerning both local and line-haul networks are not yet definitely sorted out and, in any way, prudence suggests first studying requirements and possibilities, then documenting each case prior to making a decision.

DESIGN ALTERNATIVES WITH LAN

The rush to present the local architectures evidently has a commercial background. Different forecasts indicate that by 1985 the different manufacturers will be selling nodes at a rate of 50,000 to 60,000 a year; that is a big market in comparison with the total base of fewer than 10,000 nodes installed today.

Among the broadband solutions proposed by equipment manufacturers are Wangnet, of Wang Laboratories, and Alan (Advanced Local Area Network) of 3M/Interactive Systems. We have made references to both of them. Wangnet offers three bands:

- The *Wangband* enables users of Wang office information systems and other Wang small business computers to communicate over the cable via a Z-80 based cable interface unit

- The *Interconnect* band enables any non-Wang terminal to communicate with both Wang and non-Wang systems, using standard datacomm protocols

- The *Utilities* band permits the simultaneous operation of seven independent video channels able to handle a variety of additional functions, including both teleconferencing and security

The Wangnet architecture uses a standard CATV cable as the transmission medium. A dual coaxial cable link is employed: One 350-MHz cable transmits, the second cable receives information. Access can be distributed in a flexible and expandable tree structure which spans a distance of up to 7 km, depending on the specific application.

Wang has also announced an "Integrated Information Systems" (IIS) network capable of carrying out both data and word processing. IIS comes in two versions: one for the predominantly data processing applications, the other for word processing. The former is handled by the VS line of Wang computers, the latter by the OIS (Office Information Systems).

Wang Laboratories has also developed "Wangnets": Mailway and Wise. Mailway,* an internal electronic mail system, is compatible with

---

* See also D. N. Chorafas, *Office Automation: The Productivity Challenge* (Englewood Cliffs, N.J.: Prentice-Hall, Inc., 1982).

OIS. Wise (Wang Inter-System Exchange) allows OIS systems within a building to be connected by a coaxial cable: up to 96 workstations can access each other or printers, disk storage units, and other peripherals. Such devices can be located at cable lengths of up to 800 m from each other. The data transfer rate is up to 4 Mbps.

The company is also actively pursuing development of a digitally encoded broadband local network to use only cable and provide multiple services for each department. This capitalizes on the lessons learned from the cable TV, its manufacturing concepts, and economics. Broadband cabling can be upward-expandable from individual departments to total systems with some 5000 stations.

Datapoint's contribution to local networking is the attached resource computer (ARC) system. It makes feasible the interconnection of job entry stations and the incorporation of gateways to remote network facilities. Two versions are available: System 40 (of particular interest to large data centers) and the recently introduced System 20, which provides low-cost communications for low-throughput devices. Both systems are based on the CSMA/CD packet communications approach.

This local network offers transparent protocol implementation and services, including packet assembly/disassembly, buffering, virtual circuit control management, data security, protocol, and code conversion. ARC provides access to a broadband medium through a software-controlled transceiver with a variety of host interfaces.

The Chase Manhattan Bank has set up ARC local networks in London, Singapore, and Tokyo which feed into the New York data center via satellite. There are several other users, and this network architecture will probably have a good future, as Datapoint is one of the leaders in office automation.

AT&T's local networking approach is based on the premise that users should consider data communications as an extension of voice communications. This approach leaves local chores to the user public, but with central control in the hands of the carrier. Bell may construct the network so that its eventual employment of satellites and fiber optic links will be transparent to the user, providing the user with modems, installation, and service. The user will only need to make sure that his or her input/output devices have a correct interface to the right modem.

Bell Laboratories seems to have examined the local network question carefully. The forementioned strategy aside, AT&T will have to be able to interconnect with locally installed user facilities for long-haul and network-to-network communications. Since the PBXs are the gateways to the overall carrier system, AT&T must follow closely developments in local networking; it cannot restrict itself just to offering the PBX.

Northern Telecom has developed a local networking system called "Omnilink"; it enables users to link their units together with shared disk files and printers. The processors can be connected to 1.7 km apart by a baseband coaxial cable, eight processors in a single net. Omnilink installations communicate with other Northern Telecom systems or with other vendors' mainframes via a library of protocols: asynchronous, bisync, and SDLC.

Prime Computer has made available "Primenet," which offers the possibility of operating up to 15 Prime computer systems in a local ring network. The distance between any two nodes can be up to 250 m. Primenet is capable of communicating with any other manufacturer's system that uses the X.25 standard. It currently supports Telenet and Tymnet in the United States, and a number of overseas systems.

Prime computers using Primenet are able to interface with mainframes from other manufacturers. Remote job entry (RJE) emulation packages are available for CDC, Honeywell, IBM, ICL, and Univac computers.

"Mitrenet" is a local network implemented by Mitre Corporation in which the 300-MHz bandwidth of cabled television (CATV) is divided by FDM into voice, data, and video channels. Within certain of these channels, individual devices share the capacity by means of either TDM or contention control.

The "Z-Net" system architecture, an expandable local network, was introduced by Zilog in 1980. The company anticipates announcing a gateway technology that will permit remote Z-Nets to intercommunicate. With the current release, the applications are electronic mail, office automation, and a commercial small business environment such as inventory control and order entry. Up to 255 stations can be connected to a maximum of 2000 m of TV-type coaxial cable, using standard connectors. Data transmission is at rates up to 800 kbps. The principle is distributing a system's low-cost elements (such as CPU, memory, and micro) while sharing the high-cost ones (peripherals and database).

A. B. Dick has introduced the "Magna III Information Processing System," which can be used as a standalone text processor or as the backbone of an expanded office automation system. It is based on a loop concept with a high-speed synchronous protocol using a two-conductor cable to connect up to 255 stations. Operators can share disk files and other equipment, such as printers. Future additions include new software packages handling advanced communications, mass storage devices, intelligent copiers, and optical character readers, but the current release supports workstations with a keyboard, a 20-line CRT, and a standard 55-character per second printer, the operating software being included with machine purchase.

Other local network systems are Ungermann-Bass' "Net/One" and

Nestar Systems' "Cluster/One." Cluster/One Model A (A stands for Apple) was introduced in the second quarter of 1980. It enables up to 65 Apple II computers to be linked together, the medium being a 16 wire ribbon cable.

Net/One, uses various types of coaxial cable, including CATV. Without repeaters, the local network extends 1.3 km, with an additional 1.3 km added with the help of a repeater now under development. As many as 250 nodes can be attached to the basic length, and Net/One is said to be almost entirely vendor independent in terms of the hardware that can be connected to it. Each node is programmable, either locally or remotely.

Cluster/One is a flexible, modular system which can be put in operation with as few as two workstations. Like the other local area networks, it features

1. A *communications server* through which it interconnects to other networks, minis, and mainframes

Let's notice the File Transfer Protocol (FTP), for another Cluster/One has been developed by Nestar, but independent producers have contributed other communications software such as Linkline and Owllink which emulates known IBM-compatible synchronous and asynchronous protocols.

2. A *File Server* offering a range of storage capabilities, from an entry-level 1.2 megabyte floppy disk to over 4 gigabyte of hard disk storage (using multiple file servers).

It is however a good policy dictated by experience to set up a different LAN everytime another file server is employed. The cost of the flat wire is minimal, and database contention problems are avoided.

Also part of the local area storage facility is a high-speed tape backup cartridge. This allows up to 2 MByte of hard disk storage on a single tape cartridge to be backed up in less than 12 minutes.

3. A *print server* able to support spooled printing; this makes feasible lean workstations with softcopy (video) output, concentrating the hardcopy requirements on a LAN level

A number of implementation packages are available, including offerings by Nestar and by independent suppliers. Among the former is the *Messenger*, an easy to use teletex (electronic mail) program integrating standard office chores and enhancing interoffice communications. An example of the latter is Interactive Videotex with color and graphics (provided by Owl, a British software firm).

Among other examples, Cromemco's C-Net is promoted for use in offices, factories, laboratories, and educational institutions. It employs shielded twisted-pair wire configured as a bus network. Its tree trunk and branch method of connecting the stations resemble other LAN offerings.

A particularity (on which I am not convinced) is that the C-Net interface will allow five to nine users to communicate over a single shared network node. A single C-Net cable span can support up to 255 users, with workstations separated by as much as 2100 m along a single cable span.

Destek's Desnet employs a bus topology, with an aggregate data rate of 2 Mbps. The Desnet uses an RS-232C serial port or a Centronics-compatible parallel port. This LAN supports bisynchronous and asynchronous protocols, and is said to be adaptable to almost any transmission medium, including baseband and broadband, via a modem. Three different network interface boards are available, one for:

- the S-100 bus type systems
- the IBM Personal Computers
- the Multibus systems

This local network can connect combinations of up to nine unlike processors and one Centronics printer at the same location.

Hinet by Digital Microsystems claims to have over 500 networks installed, each linking from 2 to 32 network stations along a single cable. Hinet is a distributed microprocessing master/slave polled network with 500 kbps twisted-pair wire for all network functions.

Designwise, each network station features a Z-80A, 64K microcomputer in configurations including standalone terminals with networking capability and basic user stations. Network stations can employ hard disk storage, floppy disk, or streamer-tape backup.

Like many other LAN, Hinet permits the use of shared peripherals, both for data storage (Winchester hard disks) and for printing through series and parallel ports. Print spooling features and the proprietary intelligent hard disk controller permit all network stations to access the peripherals.

Redundant master capability allows users with real-time data security requirements to have a second redundant master station to continuously and automatically backup all disk writers. Should the network's master station fail, the backup master station comes online with little interruption of the local network's operation.

Another commercially available token access ring network is Polynet, based on the original Cambridge Ring, which initiated prototype operation in 1980. Polynet was released by Logica; it uses twisted-pair telephone cable, at rates up to 10 mbps.

The Cambridge Ring was born from a proposal of Cambridge University computer pioneer, Professor Maurice Wilkes. The idea behind the system is to connect all the stations of a local network into a single-channel ring by means of serially connected shift registers contained in each station's interface. By synchronously clocking all these shift registers (say at a data rate of 10 Mbps), a fixed length bit train can be made to travel continuously around the conceptional division into a fixed number of slots, each 38 bits long. When a station wishes to transmit over the network, it merely monitors the circulating bit stream until it detects an empty slot, which it then replaces with a mini packet. Destination stations monitor the bit stream until they detect a mini packet bearing their own address.

Corvus announced Constellation, a star-type cluster system in March 1980. It links up to 64 microcomputers, allowing them to share 5 to 80 megabytes of mass storage and other peripherals such as printers and modems. To data, Corvus says more than 2000 Multiplexer networks have been installed worldwide.

In 1981, Corvus Systems introduced Omninet, a LAN for microcomputers, with the ability to link as many as 64 of the same or different brand computers (provided they conform to CP/M and the S-100 bus). These machines can communicate interactively with each other and with peripherals at speeds up to 1 Mbps, over a 1300-m twisted-pair serial link.

Each workstation attached to Omninet has an interface controller, or transporter. It interfaces directly to both the twisted pair and the host computer's memory. The data transfer is handled by the transporter and data buffering is not required. Hence, high-speed data transfers are possible while maintaining a low-cost network interface.

No master controller is required on Omninet (and with many other LAN). Network control is assumed by any transporter that has a message to send as soon as the network is available. This and similar solutions allow any device to transmit at any time with a collision-avoidance scheme of the CSMA type being implemented.

Contrary to Ethernet and Cluster/One, Omninet does not have a collision-detection mechanism. Instead, each workstation waits a random amount of time before transmitting. This eliminates the cost associated with collision-detection hardware. Also, it makes feasible protocol implementation over longer distances since propagation delays have a lesser effect on collisions.

Finally, it is also proper to mention the PBX-based central data switches as an alternative to the LAN. They feature kilobit data rates and their manufacturers underline the advantage of using the existing telephone wiring.

Vendors of PBX local networking include Intecom Inc., Northern Telecom, Rolm, AT&T, Datapoint, IBM, and Honeywell. What companies promoting the PBX approach are not always saying, is that the PBX will provide telephone service and *limited* data/text terminal service. Through the latter, it may accommodate word and data processing at desk level, as well as access to a mainframe computer.

Chapter 19

NETWORK SUPPORTED SERVICES

INTRODUCTION

Local and long-haul networks have made it feasible for many organizations to realize their objectives. These include using computers without special operators, letting the end users run the system, coinvolving the remote end user at his or her workplace, operating autonomy, and enjoying significant reliability in both equipment and operators.

Good network architectures are characterized by an ease of implementation; assure flexibility and expandability; provide a direct line of communications, workstation to mini and mainframe; permit efficient administration of local databases and online access to the central files; and by reaching every workplace, facilitate training to assure management's comprehension. Furthermore, because they are modular, distributed information systems enhance the ability to build a computerized organization. (Because of the vital role of information technology, it is foreseen that in the 1980s, the rate of bankruptcies will be higher for companies unable to adapt to and use information systems to their advantage.)

These references are just as valid for data processing, databasing, datacomm, in the now established classic sense—and for newer fields such as CAD/CAM, robotics, and office automation. Whether we talk of text or data, there is no fundamental difference in the manner of approaching organizational analysis and systems design for networking. The difference is between doing a job well, doing it wrong, and not doing it at all. The implementation of the third generation of online systems has prerequisites and the development of an efficient long-haul network is one of the milestones.

Not all companies have the financial resources and knowledge to build such a network and they are, therefore, looking toward shared, switched data communications network service with value-added characteristics (store and forward, error detection and correction, change of speed and code, and so on). Such a network should typically provide for:

- Sharing communications facilities
- Interfacing incompatible terminals and computers
- Offering a wide range of data communications capabilities from which users can select as needed
- Managing and reporting network performance

Compatibility is a major problem for firms that have implemented various systems piecemeal. Organizations attempting to develop a distributed processing system cannot always use the most suitable equipment because of interfacing. Often, the equipment used in different systems was selected without regard to compatibility, with the result that applications cannot be integrated into a more efficient common system. This applies even more strongly with respect to equipment of different origins.

The shopping list of the facilities that users would like to see supported by a value-added service can get long. The time is fast approaching when tapping a real-time global database would make the difference between success and failure for a firm. The same is true of impressing management with a virtually instantaneous readout of facts and figures.

The competitive edge these days relates to online access. To provide that access conveniently, efficiently, and economically, organizations are discovering the virtues of networking. Through the services supported by a modern network, they can put computer power where it is needed most: right where the action is.

BELL'S VALUE-ADDED NETWORK

As stated in Chapter 16, AT&T's upcoming answer to public packet networks is a nationwide network handling various protocols. The Advanced Information System/Net 1* includes communications process-

---

* AT&T has also introduced a packet-switching service known as BPS.

ing and network management capabilities and, as stated, will compete with the packet networks operated by GTE-Telenet, Tymnet, IBM's Information Network, and other VAN organizations.*

Communications services to be provided by Net 1 include communications processing functions such as data entry, inquiry/response, remote job entry, and message distribution. These functions can be customized by customers using a high-level programming language. Implementation of the processing functions will be provided for in storage areas of the ACS nodes. Each node will have several general-purpose storage units for data accumulation and a processor to execute a user's instruction set.

Net 1 will include as an integral part of its structure many of the communications-oriented functions now being performed by the customer's on-premises hardware and software. The network will connect terminals to terminals, terminals to hosts, or hosts to hosts on a 24-hour per day, seven-day per week basis. Standard and priority service grades will be provided for transmission.

Each grade of service has several delivery options. Once a user selects the level desired, billing is computed on the basis of transmission volume, without regard for the amount of transmission time or the distance traversed.

Information released on Net 1 and its predecessor on the drawing board ACS (Advanced Communications Service) states that within the priority service designation, three levels can be selected: Priority 1 assures end-to-end transmission in a maximum of 200 ms; Priority 2 brings the threshold to 15 seconds; and Priority 3 extends transmission to a maximum of 30 minutes. The standard grade of service will provide for end-to-end transmission in a maximum of four hours, and this may be sufficient for a large number of applications.

Within the four service grades, three transmission delivery options will be available:

- Standard
- Delivery when authorized
- Timed delivery

Standard delivery means that data are transmitted directly to the addressed terminal or host. If the destination device is busy or offline, the

---

* Prior to the constitution of American Bell, AT&T had in the works a Value Added Network, the Advanced Communications Service, which was never brought to the market, with the exception of the so-called ACS II test-marketed in New York City. Under American Bell, ASC II became Advanced Information System/ Net 1—until lawyers by Ungermann-Bass protested that this vendor had an offering of the same name. The name was changed to Net 1000 after this book was typeset; so it has not been possible to update the name throughout.

message is held at the serving node, with attempts to deliver made at 1-second intervals.

The delivery when authorized option sees to it that messages are sent to the destination node and held in the user's general-purpose storage unit until the terminal or host requests them. Under the timed delivery option, data are delivered at the destination node at the time of day specified by the user.

Terminals and computers using AIS facilities will terminate at a local communications node port. Such node ports, according to the AT&T specifications, will be similar in function to ports found on communications front end processors. The service will include access charges that cover transmission between the user's premises and the node.

Terminations at a node will depend on the equipment the user has attached to the access line. Port accesses at a node will be configured to support specific digital communications protocols, equipment, and in some cases, specific software packages. The network emulates host processor acknowledgments of received transmissions. This signals that the data are properly received by the network and not by the host or terminal.

Design specifications maintain that users will be able to mix equipment on the network, with ports designed to accept each type of terminal in the user's net. It is, however, most likely that this will come in phases, with some equipment accepted earlier than others.

Several types of device control are described in the specifications. IBM asynchronous *EBCDIC* line control will be, for instance, available to support the 2740 Model 2 and other compatible equipment. The access arrangements will be provided under EBCDIC line control: dedicated, outward dial, and shared usage.

Dedicated access can include 2 to 20 terminals as long as the terminals are all on the same premises. This type of access includes a dedicated private line between the node and the user's premises.

Outward dial access uses the dial-up phone network to access a terminal with AIS-compatible autoanswer features. The packet service will provide both dedicated and shared dial access ports. The third access arrangement available under the EBCDIC line control will be a shared, multiuser, multidrop line. This arrangement will be limited to Priority 3 and standard grades of transmission because of multiple users sharing a single ACS node.

The EBCDIC line control will support terminals operating at 300, 600, 1200, and 2000 bps. Terminals with buffer storage ranging from 512 to 5000 characters will be accepted, but the user must specify the terminal speed and buffer size.

ACS will also provide the asynchronous Teletype line control used in the TTY Model 33 and the Model 35 terminals. The Teletype line control will provide half-duplex asynchronous transmission, ASCII code, with dedicated dial access of 110 and 1200 bps. With this offering, only dial-up access ports will be available, so terminals will have to be equipped with compatible autoanswer dial capabilities, according to the specifications used. The binary synchronous (BSC) access is a third method.

THE EMPHASIS IS ON COMPETITION

The AIS network will allow users to define and tailor their business needs into a range of supported communications services; it will also provide common user software communications packages for a monthly fee. These packages will be written in a standard language and can be modified by the users, but user-modified packages will have to be submitted to the Net 1 Customer Service Support Center for validation. This center will test all customer-written or altered programs to ensure that they do not affect other users of network operations.

As the information available today indicates, the four packaged services available on AIS will be data entry, transaction, storage administration, and terminal/host addressing. However, as has been the case with IBM's SNA and all other system architectures, Net 1 can be expected to expand its capabilities and deliver new services and operating efficiencies over the next years following its implementation. In addition, Bell has developed a microwave radio system that reportedly will handle three times as many data and voice calls as its current top-capacity long-distance radio link.

The communications network will involve electronic switching systems (ESSs) interconnected by a high-speed signaling system to enable more effective operations. The combination of ESS and the signaling system brings to bear the stored program controlled (SPC) network. As it evolves, users will notice faster setup times for call connections, as fast as 2 seconds for a coast-to-coast call.

A complementary part of the effort will be expansion of the new signaling system, known as Common Channel Interoffice Signaling (CCIS). Other possibilities for further services include:

- Identification of priority callers by means of distinctive ringing for calls from preselected numbers

- An automatic callback system that would make note of a busy signal, wait until the line was clear, and then dial both numbers and connect the two lines

- Preselection of the telephone numbers from which collect calls could be accepted without operator assistance

This will eventually interconnect all the new ESSs, enabling them to share call-handling information in high-speed digital form over communications pathways separate from voice circuits. Some of the new services reportedly will stem from the ability of the SPC facility to forward a caller's telephone number to the local central office serving the called party. The SPC facility will also offer other services designed specifically for business customers.

AT&T has also developed an advanced microwave system said to handle three times as many data and voice calls as its current top-capacity long-distance radio link. The technology of this AR6A system permits more efficient use of the radio-frequency spectrum now available for long-distance telecommunications.

Bell's AR6A will be the first of its kind to use single-sideband (SSB) technology for high-capacity, long-distance microwave transmission, and will be able to carry over 6000 calls per channel, compared to 1800 for Bell's highest-capacity microwave radio link (TH-3) now used.

SSB packs more telephone conversations into a radio channel without sacrificing transmission quality. As a result, a greater capacity can be carried in a frequency band and the frequency spectrum is used more efficiently. AR6A operates at a frequency of 6 GHz and handles both voice and data signals. By the late 1980s, SSB systems will be available for most of the new long-distance mileage added to the Bell System's microwave network.

With all its technological innovations, AIS/Net 1 is one of the many networks operating or in development in the United States (Telenet, Tymnet, and others), Canada (Datapac), France (Transpac), Germany (Datex-N, Datex-P), Scandinavia (NPDN), England (PSS), South Africa (Saponet), Spain (CTNL), and Japan (DDX, ISDN). These developments indicate a common objective to provide more efficient and economical data communications services through new, high-quality public networks. Many of these new digital aggregates employ the latest transmission and switching technology, not only for data communications but also for text, image, and eventually voice.

A variety of services is available in these new networks, including leased circuit-switching and packet-switching operation, but digital systems impose new demands on protocols. The transmit and receive circuits convey both control information and data, depending on the operating phase. The control circuit serves the basic hook-on function to start the call establishment phase or the clearing phase. A key feature is to maintain transparency, because it prevents data from talking a circuit into a disconnect, or misoperation, with certain bit patterns.

Based on South Africa's Saponet experience, Figure 19-1 presents the Transport Activity (TA) as the channel between the network and the DTE. This transport activity involves:

1. Message segmentation at packet transport level
2. Message segmentation at logical terminal level (which can be a physical terminal, or processes within a physical DTE, database, or buffer)
3. Virtual circuit management
4. Routing
5. Transit control

Within the context of this responsibility, the long-haul network transports node to node. To the nodes are attached programmable terminal controllers and to the terminal controllers are connected the DTE, hosts, databases, and terminal concentrators (Figure 19-2).

As Olivetti (the designer of Saponet) improved on the original South African experience, the PTC can come in two versions: a more limited PTC I, which provides network interface, terminal control,

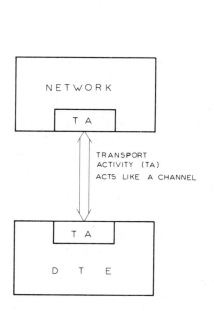

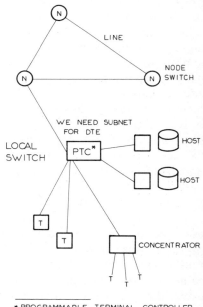

FIGURE 19-1   The transport activity as the channel between the network and the data-terminating equipment.

FIGURE 19-2   Functions of a programmable terminal controller acting as a local switch.

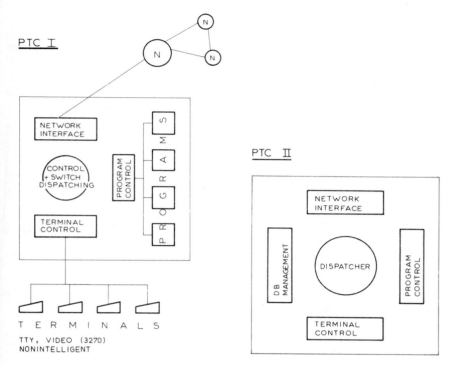

FIGURE 19-3   Two options of a programmable terminal controller: with and without database management capability.

stored program capability, and dispatching functions, and a more sophisticated one (PTC II), which also supports database management. The functional elements of PTC I and PTC II are shown in Figure 19-3.

SHARING THE AVAILABLE FACILITIES

The introduction placed due emphasis on facilities sharing; this and compatibility provide for a nationwide communications facility that can serve data communications needs in a manner similar to the way the phone system serves voice communications.

With value-added networks landing in all cities where a company operates, network architecture and protocols are taken out of the hands of computer vendors, whose primary concern is the sale of their own hardware, and placed where it should be, in a service furnishing a shared, compatible communications system. Compatibility in communications means generalized protocols. Security/protection requires the support of closed user groups (CUG).

A closed user group capability enables the user organization to utilize shared facilities *as if* they were its own private network. Through trap words and the assignment of codes, the carrier is able to guarantee that no unauthorized eyes will see the traffic concerning a given CUG or reach a CUG's own databases. The British Prestel system (in operation since March 1979) has been one of the first widespread text and data networks to offer the CUG facility. In spite of some 8000 terminals currently operating on Prestel and installed at the most diverse places, there has been no known CUG violation so far.

In terms of compatibility, initially AIS will support protocols for what AT&T considers to be more than two-thirds of presently installed general-purpose terminals (teleprinters, CRTs, and remote batch devices). Special-purpose terminals such as plotters and facsimile transceivers will be supported at a later date.

Various disciplines will be provided for interfacing Net 1 with host computers:

1. Emulation support, initially for ease of user entry to AIS

Emulation support makes the AIS network look to the host as if it were interfacing directly with character-oriented terminals.

2. Enhanced capabilities that free the application programs from terminal dependence

Referred to as *message-level interfaces*, these perform functions similar to those of communications software such as IBM's BTAM.

3. Communications modes available to users, corresponding to on-line and batch-type operations

Such communications modes will support the call feature for interactive processing, such as inquiry/response and timesharing, and the message feature, intended for such one-way transmission tasks as store and forward, remote batch, and data entry.

The call feature provides priority levels for users to trade off cost against speed of delivery, a subject we have already considered. The call feature sees through terminals and computers in operations between one another in a bidirectional, fast-response mode. Thus terminal-to-terminal communications will be immediate, just as telephone conversations now are.

In this manner, terminal calls can economically replace phone calls when the incremental cost is low because the network is there for other purposes anyway. Calls can be manipulated by users in various ways:

- Place a call on hold at the terminal or at the destination.
- Receive a call.

- Forward a call to another station.
- Authorize the receipt of collect calls.

The sender can build up messages, edit them, add to them, and designate the time of transmission and grade of service. (We have made reference to different grades of service: priority, standard, deferred.) In addition, messages will be journaled at the send and receive locations.

IS THE VAN AN ALTERNATIVE
TO A SATELLITE SERVICE?

To answer this question, we will consider a specific example. American Bell's Net 1 and SBS are different businesses. Although together IBM and AT&T are likely to represent the largest share of the total network market, in SBS IBM is a partner with Aetna Insurance and Comsat. This is a difference in ownership and it is just the beginning of the reference.

The SBS system features specialized common carrier rooftop antennas that will allow establishing point-to-point communications on demand without using terrestrial facilities. It is purely a communications facility, analogous to the WATS service provided by the telephone company, with a rumored goal of 25% return on investment and 25% of the potential market.

A basic objective in SBS is to provide a low-cost alternative to leasing lines from AT&T. High communications line charges now limit the growth of distributed information systems, and IBM would like to eliminate this constraint on its market. In fact, any development that reduces the cost of data communications will in the long run benefit all suppliers of data processing equipment: mainframes, minicomputers, terminals, and peripheral equipment.

As a Value Added Network, AIS/Net 1 is a complete network architecture and compares with IBM's Information Network (not SBS), but it also provides two basic classes of transmission service, *call* and *message*, which other VAN do not necessarily support. The call service gives the customer a two-way transmission path between originating and terminating stations and is oriented toward inquiry/response systems. The message service assures a range of functions for the presentation and handling of messages and for directing their movement through the network. Data entry and remote batch are examples.

Other technical differences exist. SBS acts as a carrier, but there is no known full range of software-based services in its offering. It is the other way around with AIS, and although AT&T announced that its network will accommodate equipment from any vendor, the prevailing thinking is that mainframes attached to Net 1 will have to be reprogrammed or front-ended to accommodate AIS standards.

The Advanced Information Systems features may not be compatible with those supported by announced network architectures—local or remote. In a way, if a user selects AIS, he or she will forgo the functions provided by the mainframe system vendors in their architecturally structured networks, such as, for instance, IBM's SNA. This can work two ways.

Since many SNA features are attractive to the large Fortune 500-type user, the large user will tend to stay in the mainstream of IBM (or other system vendor) system development, implementing SNA-based or other architectural systems rather than AIS-oriented approaches. With program portability being increasingly assured at the computer level, diverse network architectures will probably be the real wall of incompatibility in the 1980s.

However, Net 1 will be an attractive option for the medium-size and small user organization whose costs of implementing an SNA-based system outweigh the benefits and for whom the issue of network control is less significant. At the computer level, the mainframe system vendors will also undoubtedly provide an interface to AIS just as they have done for the current packet-switching networks using the CCITT X.25 standard.

In other words, any particular hardware manufacturer that wants to take advantage of the relatively low cost service available over AIS will have to adapt its hardware to the network standards established by AT&T. This is no different from the problem facing the plug-compatible peripheral manufacturers today in attaching their equipment to IBM computer systems. It is a technical problem that can certainly be overcome.

The possibility is also present that as the price of raw hardware declines and the software necessary to make the systems work becomes a more important source of revenue, the attractiveness of the plug-compatible strategy declines both in revenue potential and in terms of account control. This, if it happens, will tend to alter the way in which users and vendors look at compatibility and may bring about new competition.

Since account control through the establishment of unique systems standards has been a characteristic of computer systems marketing for some time (and is a practice not without technical justification), the process we have been describing will represent a difficult choice for the systems vendors to make. Account control is, in fact, moving at the network level of implementation.

Finally, technical challenges to one side, network-level implementation will be faced with transborder data flow problems, which have haunted the multinationals over the last couple of years. As we will see in the following chapter, transborder data flow has become a subject of legislation in many countries. It will be magnified even more by the fact that many developing countries want to put communications ahead of ideology.

CONNECTING TO THE OFFICE ENVIRONMENT

While the preceding sections have focused on long-haul Value Added Networks, it will be good to remember the VAN's extension into the office area which, in the mid to late 1980s will be typically based on local area networks—and not on mainframes or minis. Forward-looking companies have found that out and are taking the right initiatives with LAN installations, as the following references help document.

Atlantic Richfield has installed a pilot project that links 50 employees in its Los Angeles corporate headquarters. As the director of Arco's office support program suggested, "People are collaborating more on documents; it's affecting the working style." At the Transamerica Corporation offices, some 200 employees on three floors are connected by a network and the user organization is looking ahead to improved quality in the way people work.

The Minnesota-based St. Paul Insurance Companies have 47 ARC systems at various locations around the United States. At each remote location, there are some 5 to 15 terminals, one or two printers, a disk server, and a communications processor.

This insurance group uses the remote installations for data entry and remote printing, having converted in 1980 to ARC from older Datapoint equipment. At the present time these ARC systems are not interconnected, except in the sense that they are all polled daily by an IBM 370. But the application is an excellent example of what can be achieved with LAN and VAN.

If ARC is the oldest, and therefore most reasonably mature local area network in the market, another vendor who tries hard is Xerox. The office equipment manufacturer encouraged both Intel and Digital Equipment to make computers that can run on Ethernet, and has sold licenses for a nominal $1000 to some 100 companies to build compatible equipment. As a senior Xerox executive was to remark, "We don't sell Ethernet. We sell office systems, and Ethernet is part of the office system."

Let's then take this example to look a little more carefully into what constitutes a local area network's contribution in an office environment and what can be achieved through LAN usage. (This reference is no claim to Ethernet's superiority over competing schemes. In fact, until Intel or some other vendor comes up with chip-level interfaces and brings the cost of the connection below the competing baseband alternatives, I consider Ethernet the lesser choice.)

As discussed in Chapter 17, *Ethernet* is a distributed multiprocessing network first developed at the Xerox Palo Alto Research Center and based on Aloha (contention) principles, but not sliced Aloha. There are other differences: Aloha uses a star topology and radio transmission, whereas Ethernet:

- Uses coaxial cable,

- Has no central control (distributed stations),

- Features a 10-μs delay,

- Provides carrier protection (testing the channel to assure that it is idle),

- Has made the interval adaptive to transmission requirements (the increasing channel stability), and

- Has collision detection, but

- There is no sequencing given that datagram solution (not virtual circuit) is employed.

Ethernet has been one of the first passive coaxial cable networks for local networking. The prototype system has been operating in Palo Alto since 1973. Other test locations include the White House, the House of Representatives, and several universities. The Xerox/DEC/Intel consortium now offers a modified version as a commercial product.

The *Ether* is a coaxial cable now, but it can also be composed of optical fibers. An originally supported 3-Mbps capacity has been increased to 10 Mbps. There are no nodes in the network, because the node function is given to gateways which detect the DTE which is down and delete it from the active list. (In the original version, all data-terminating equipment was computers; no terminal was directly connected to Ether, which explains and partly justifies the high interconnection costs.)

Decision on structure and topology has considered four alternatives: Should the local network be fully connected? The designers answered the question negatively because the resulting system may be too expensive. Should it be star? No! It's an old concept. Store and forward (Arpanet type)? Neither! It does not need nodes, as only computers are connected. Should it be ring? No! Rather the Ether structure, as chosen.

Ether transceivers insert and extract bits of information as they pass by on the cable; controllers enable all sorts of equipment, from communicating typewriters to mainframe computers, regardless of the manufacturer, to connect to the Ethernet; and the use of software answers a key problem in any system of this type: how to control access to the cable. The local network architecture sees to it that while one component transmits, everyone else waits; when it stops, those who want to transmit do so. If messages collide with each other, the transmitters stop and wait to start up again.

Still, Ethernet is a passive communications network with no switching logic. It simply accepts transmissions from attached system elements, a technique providing reasonable simplicity and, therefore, reliability.

The coaxial cable core of Ethernet is made up of one or more segments, typically up to 500 m long, and communications transceivers, one for each element of the office system. The transmitted packets are of datagram type, and feature arbitrary length with a 32-kilobyte practical limit, owing basically to error rates and buffer constraints.

The network connects system elements within a building or part of it. Other transceivers with associated processors could be used to connect different Ethernet to each other and to outside communications facilities for long-haul transmission. Each element has a unique address. Information is transferred in packets, which include:

- Data to be sent
- Address of the unit that will receive
- Address of the unit sending

Each transceiver monitors the network before transmission to be sure that it is clear and during transmission to detect interference. If there is interference, the packet is resent later. To receive data, each element recognizes messages with its own address, accepting those and ignoring the others. When a transmission is completed, the receiving station sends a message of acknowledgment to the sending one.

Such services fall within the announced goal—to allow many users in a given organization to access the same data, obtaining economies that come from resource sharing but also being flexible, allowing users to change components easily so that the network can grow smoothly as new capability is needed. Furthermore, a system based on the notion of shared information needs reliability.

Given its structure, the Ether can be extended from any point just by adding wire. For internetworking purposes, an Ethernet has a unique address which communications between networks can use. Once a packet has passed from one network to another, it is handled in the same way as one sent locally. Messages can also be sent to and from non-Ethernet networks through gateways that monitor the data exchange.

There are also problems. Some relate to the fact that a broadcast base has been adopted for transmission. Although this has many strong points, the nature of a broadcasting system is that almost anybody can tap in and read the data; security tends to be low. To guarantee secure communications between any two particular hosts, it is necessary to implement some form of encryption, an added expense.

With the current release, the connected DTEs make a copy as a packet passes by; source and destination are both detected. The address indicates *on us* or *not on us* entities. The packets are self-clocking. Retransmission takes place if needed, and it can be initiated immediately.

Ethernet backs off on retransmission, with successive collisions. Design criteria are:

- Deference (collision detection effect)
- Short contention–long acquisition
- Overall efficiency

Yet packets can be lost in case of an undetected collision, impulse noise, inactive receivers, and so on. To recapitulate, the current solution accepts 256 stations, 10-Mbps capacity, and a distributed packet-switching base with broadcast address recognition. The error rate varies with the traffic patterns, but seems to be at the level of 10^{-3} packet.

Among the software services supplied with Ethernet are: a *clearinghouse* which provides user name to network address correspondence, thus insulating the user from the need to understand the topology of the network. Another service is internetworking-routing, encapsulating Ethernet packets within the transmission protocols of common carrier facilities. This permits distantly located Ethernets to be logically tightly connected, and constitutes another issue where LAN and VAN come together.

An appropriately designed gateway translates the low- and medium-level protocols of foreign networks into the corresponding Ethernet protocols. This way, dissimilar network types can be interconnected. Furthermore, an interactive terminal service permits non-Ethernet workstations to access filing, printing, electronic mail, and other services available on this LAN.

Finally, I have also mentioned work in process to produce a local architecture for optical fiber media. The Fibernet design differs from Ethernet in two important ways: the insertion loss from connectors and splices in an optic fiber is still quite high; reflections at connectors and splices are also considerable. Hence Fibernet employs shorter transmission lengths and uses a passive transmissive star coupler. Two fibers per host are currently employed, as transmission must be unidirectional. Since transmission is unidirectional, monitoring for collisions remains possible despite reflections inside the system. There are no known plans for a commercial announcement of Fibernet.

Chapter 20
PRIVACY

INTRODUCTION

The major issue of the "Information Revolution" to confront us now and in the future is not technology, but its effects on our social life. All aspects of our social lives will be touched, but most particularly the intricate mechanism needed to support personal freedom.

Computers and communications—voice, image, and data integration on one network; packet-switching, satellite-supported broadband capabilities—will cause great social change during the 1980s. Add to it viewdata and personal computers and you extend the far-reaching ability of the communications technologies into every home and office. We need legislation specifically designed for individual privacy.

Restrictions on the use of records, whether they concern medical, scholastic, employment, military, or government information, must be explicit. Privacy laws concerning banking, credit, taxation, and insurance must be reexamined in terms of online access, public and private databases, and all issues violating personal privacy.

Both the law and law enforcement should be realistic. There is no absolute safeguard against unlawful access to computer-stored information, just as there is no absolute safeguard against the unlawful use of an automobile. But there is a system of lights and signs to regulate the flow of motor traffic and there should be a similar control system to regulate the flow of information.

Not just federal agencies, but all organizations that use computers to manage files, detect fraud, or search for abuse must abide by regulations. Safeguards for all sorts of records should limit government access to them, give individuals the right to see what is written in files concerning them, and make it a crime to obtain record information under false pretenses.

These are but a few examples of the privacy protection mechanism that must be established. Legislation on the protection of records should provide a legal standard of confidentiality. It should also require an organization to tell its employees that information about them might be used and provide criminal fines for unauthorized disclosure of personal data.

Both industry and various government departments should allow persons to know what information is being collected about them and to see and correct their personal files. This is a practice already voted into law in the United States and in other countries. Individuals should be informed when an adverse decision is based on personal data and be allowed to prevent improper access to their files.

All this must be safeguarded by law as databases spread around and communications media make feasible increasingly easy access. Personal computers and viewdata services compound the problem, and the increasing complexity and sensitivity of information maintained in an employee's personal records poses a mounting threat to one's right of privacy.

The potential for harm to a person has increased dramatically because so much personal information is compiled in different places, accessible online, and transferable any place upon request. "Voluntary compliance" to poorly specified standards in personal records' privacy is no protection at all.

Protection is effective when an individual is legally permitted to inspect computer-based documents that directly affect his or her status, and there should be no arbitrary exception of documents.

The protection of freedom is guaranteed when formal means exist for individuals to point to and request correction of errors in the records they have inspected, provided that they can present the needed documentation. Furthermore, the opportunity to inspect and obtain corrections of one's record should be clearly available and communicated. And this is just as true for verification that the documented change has been made.

A NEED FOR LEGISLATION

How far the protection of privacy can go in the United States and how much legal background is provided will depend in the way the American public views the issue of computer-based record keeping. Statistics are encouraging. An insurance survey proves that:

- Two out of three Americans are concerned about the privacy and confidentiality of their own records.

- The public accepts the collection of information if relevant to the purpose for which it is being collected and if it is not too intrusive.

These *ifs* leave some gaping holes. How can we tell the difference? What is legitimate today may be intrusive tomorrow by overblown databases and their integration. Let us not forget that computers made it easier to obtain information improperly: hence the need for legislation.

There are three precise and important dimensions in this reference:

- Personal/social
- Political
- Economic

Information is power. Both national laws and international agreements should regulate the flow, storage, and access of information. Because data move transborder, whether legislators allow it or not, international accords are imperative for protecting privacy and for defining the controls to be assured.

Happily, there is an awareness of that need. Moved by a fast-paced technology, many nations are beginning to develop an information policy that encompasses communications facilities, databases, and computers. Control mechanisms come in the form of *data inspectorates* working to bar:

- Certain types of transmissions
- Unwanted database developments
- Illegal location and form of data processing organizations
- Unacceptable data records

What is allowed must be legislated, then policed to protect personal privacy. To police the police is a tough task. Interpol, the international organization of national police agencies, recently faced legal problems in its plans to computerize its extensive records. [The administrator of the new data protection law in France (where Interpol is headquartered and by whose laws it must abide) intends to make Interpol justify and protect the name-linked data it passes around. Europe's data protection authorities aptly pledged support for keeping Interpol's projected computer files in conformance with privacy standards.]

In all probability, the battle which will peak in the 1980s will be over the erosion of national borders. It will range on the effect of transborder data flows on sovereignity, technological advancement, databasing, personal privacy, security, and the challenge to cultural values.

The encompassing theme of preservation and enhancement of state power will surely be brought into focus. With computers and communications networks, applications processes—not only people—can dial other processes (or people) from anywhere to anywhere, for any purpose. Who is to control this flow?

These were many background factors behind the 1977 Vienna symposium sponsored by OECD (the Organization for Economic Cooperation and Development, which followed the Marshall Plan as a cooperative effort among the United States, Canada, Western Europe, and Japan). In view of the interest the subject has aroused, government representatives, computer users, vendors, consultants, educators, lawyers, and more than 200 delegates from 20 different countries streamed into the conference center of the Vienna Redoutensaal. The inevitable consensus is that both private and public users of international data transmission and data processing will find their methods of doing business affected.

We are seeing the beginning of a worldwide effort to define what information can be gathered and transmitted, and to settle not only the content but also the methods such transmission can take. Yet, as a *Datamation* survey of 40 American multinational firms revealed, more than half are unaware of what is going in this field. One reason is that the laws are in their early stages and some organizations are just beginning to look into the subject.

OECD has begun to investigate advances in computer applications as they relate to corporate information systems and their aftermath. This is something on which OECD has done a great deal of work in the past, but also something that has been dropped as the interest became more pronounced in pure data protection issues.

The Council of Europe (COE) is also deliberating on proposals to ensure personal privacy protection in the international movement of data. The COE experts have accepted a draft convention pertaining to transborder data flow which also reflects the American viewpoint that international data protection measures should be primarily voluntary;

that the free flow of information should not be unnecessarily restricted. But what is meant by "unnecessarily"?

THE ATTITUDE OF THE AMERICAN PUBLIC

The public survey to which we made reference was presented and discussed during the National Computer Conference in New York City in June 1979. Its statistics make interesting reading, and it is worth taking a couple of pages to look at them.

* 60% of the general public sampled and
* 97% of the computer executives sampled

think that computers have improved our quality of life. Those who think that computers offer valid individual service to private persons are:

* 64% of the general public
* 86% of computer executives

In terms of negative reactions—precisely: damaging effects to individuals—the percentages are, respectively, 80 and 67%. Is there an actual threat to privacy by computer? "Yes" answered

* 54% of the general public (in 1976, only 34% gave that reply)
* 53% of computer executives

Is privacy adequately safeguarded? Here, 52% of the general public and 53% of computer executives think that we have *not* yet achieved adequate safeguards in the use of computers.

If privacy is to be protected, should computer use be sharply reduced? "Yes" said 63% of the general public—but only 8% of computer executives.

The attitude that some of the interviewed people have is that we all surrender our right to privacy the day we take out a loan or apply for a credit card. Two issues—morality in computer use and personal freedom —"better be left up to the individual rather than regulated by law" some people felt. But to the question "Are new laws needed?" the majority answered "Yes." It is important for Congress to pass legislation. And there seems to have been a general concern for organizational policy and legislation to protect the privacy of the single person.

If the position "Has technology gotten almost out of control?" (alienation index) was submitted to a national referendum, it might carry the day. In the survey 43% said yes, 41% no; and 16% were not sure. Nevertheless, the majority felt that most organizations collect too much personal information.

The implications of this type of study are left to individual determination. The trend seems to be that the public appreciates computer usage, but also fears that not enough has been done to protect personal freedom.

If effective protection is not provided, there may come a time of great scepticism about the favorable fallouts of technology, given the many and variable faces of negative fallouts. Let us not forget that technology has no morals. Humanity should have.

ALIENATION

It may be easy to point to violation of Americans' privacy via computerized data banks kept by the Internal Revenue Service, the Federal Bureau of Investigation, the Central Intelligence Agency, the National Security Agency, and other agencies, but we often tend to forget that the computers are merely tools of officials and policymakers.

There is a definite need for anticipatory analysis to meet head-on the problem of public alienation. The mechanism to be put in place must be able to relieve the fear and assure that whether we talk of personal computers, large-scale systems, or data communications, neither the network nor the single machine leapfrogs into personal privacies.

Privacy will become a keyword in the future for:

- Legislation
- Databases
- Data communications
- Organizational aspects
- Policy positions

Although it is not a research instrument but a public opinion poll, this survey revealed many things. The American public is preoccupied, and there is no reason to believe that the average person is ill-informed on the role of information in business, banking, insurance, and government bodies—from regulatory agencies to the IRS.

The public realizes that personal information has become a basic ingredient of business and banking, and this increases its concern about privacy. Yet the public must be made aware that it is difficult to assume the benefits of the producer/consumer society and also keep "private data" private.

The balancing act between potential benefit and latent risk is a difficult one. At the present time, a rumored 1 million bytes of data per person, per year, are collected in the United States alone.

In the background looms the mighty database; the trillion-byte memory to which all authorized industries from candy manufacturing to law enforcement feed in and which interrogate. In the foreground is the little, harmless, intelligent TV, viewdata station, home computer, or small business system connected online to the databases.

Online communications facilities magnify the problem, the more so as file protection, reserved access, and other measures of security do not necessarily concern the mighty agency—if the law does not explicitly say so.

Our societal complexity has far exceeded our ability to cope with it. We have created a new world (hopefully not a *Brave New World*), a world of virtual assets.

No more are the brick-and-mortar values possessed by a person the only basis of his or her wealth: data and knowledge are the new standards. Should we be afraid of them, shy away, exploit them, or regulate them?

TRANSBORDER DATA FLOW

Let us look at a precise case. Since the early 1970s, companies encountered problems in moving data from one country to the other as transborder data flow was curbed, Europe and Canada being specific references. As some countries object to information moving by leased lines, multinationals look to satellite transmission as a solution. But is it a solution?

Multinationals are beginning to show concern about what may happen to corporate communications from the evolving data protection laws. Sovereign states do not yet have a clear view, but the fear is there that local offices and data processing centers may be closed in favor of data processing at headquarters. Although laws may inhibit some data migration, many parties do realize that the problem is much deeper.

As a first answer to these challenges, eight European countries have adopted legal policies in regard to information privacy. Five new laws are forecasted for this year. Prior to the end of 1980, there were 18 countries with laws on data protection. But are these laws homogeneous? The answer is: NO.

In a number of issues, divergencies are much less evident among European countries than between them and America. The problem, though, is always the same: "What is the policy when data cross national borders?" And why only national borders? What about the borders of the household?

Since national policy is a reference, let's look at the record of the transnational data movement to provide a clue, starting perhaps with Scandinavia, where information privacy is the law of the land.

The Swedish position can be expressed as follows: *"If* there is a reason to support that a piece of information will be used for data processing purposes in a foreign country, this data cannot be communicated except only after explicit authorization."

Norwegian law foresees a special "export license" for data prior to communication to a foreign country, but the law also permits the king to enter into communications accords for international exchanges with other countries to simplify the procedure.

Denmark has adopted a less formal viewpoint on the transfer of data. Here, the law does not put into play the existence of a judicial solution between the Danish emitting organization and the receiving one in the other country. This contrasts to the Swedish policy, where specific legislation obliges the owner of private data to define the content, the responsibilities, the projected use, the security, and other fundamental characteristics.

The lack of harmony between legal texts is striking. If this divergence has not yet reached a clogging point, it is because, in terms of volume, transborder data transmission is still a small percentage of the amount of data produced in each land. But the situation is changing, and transborder data movements are gaining importance.

HOW THE CHALLENGE STARTED

The Land government of Hessen (West Germany) was the first to take the initiative of legislation whose goal was to guarantee data protection for private users (Hessisches Datenschutzgesetz). This happened in October 1970. The object of the law was to protect the rights of the individual, establishing at the same time rules for handling data in the public sector. Therefore, the law distinguishes between "data protection" and "the concept of secrecy."

Data protection has been applied particularly to the rules and legal instruments that concern the liberties and the interests of individuals. Data stored, handled, and exchanged by computer are therefore subject to the law.

Given this reference, the thesis could be advanced that the recognition that computer usage requires special legislation for data protection has developed largely in Europe. But the European concept differs in its fundamentals from the American, which considers the extent and inquisitive character of personal archives run by the government, the banks, and some other organizations as the principal menace to individual freedoms.

The American concept is fairly broader, as it defines the private domain of an individual as "his or her right to be alone." It is a person's right to conserve for himself or herself alone some personal details.

The concept is clear. But what about the mechanism for its implementation? Legislation must be precise, explicit, even detailed on the implications of databases (whether private or public). And this is still a little-known subject even to those who think they are experts.

Different levels of detail can be distinguished on each issue: from basic legislation to its applications; from the implications from databases to law enforcement and supervision (Table 20-1).

Since the value of information on an individual varies tremendously from one person (physical or legal) to another, the American concept tends to enforce the legal protection measures in terms of complaints registered by different people on the secrecy of the information concerning them. The action to be taken by the source will vary accordingly.

The distinction is tantamount to saying that legal responsibility falls on the shoulders of those individuals, government agencies, or firms

TABLE 20-1   Protection procedures

Level of Legislation
   Central government
   Local government
   The two apply different texts
   Or there is only one text valid for everybody
   Same reference for national governments and international accords

Implications to the Database
   Local computer handling only
   Computer network availability
   Organized databases
   Data heavens
   Both computer-based and manual files, but linked to one another
   Computer-based and manual files independent of one another
   Manual files only

Mechanisms for Application and Supervision
   Authority responsible for surveillance
   Protection by the judiciary
   Administration of the inspection function
   Management of the interfaces for regimentation and legislation

---

abusing the privacy of a person. Such is the sense of the Privacy Act of 1974. In simpler terms it reflects the wording of a judge's decision: "Your freedom ends where your neighbor's nose begins."

The approach of European legislation is that private interests can be protected through a number of procedures and rules to which all organizations that run personal files must conform. This answers some of the reservations expressed in the preceding paragraph, but leaves something wanting on other issues.

Can all cases be foreseen and regulated at this early stage, in front of an explosive computer and communications technology which promises to change the face of the earth and society? The answer has to be negative still, the American approach better answers this query.

THE DATA HEAVEN

In principle, the laws must be few and simple. Only then are they valuable. And they must not only protect freedoms but also assure that they do not leave major loopholes.

Such is the case of the *data heaven*, denounced by the Swedes. The risk of a data heaven fits squarely into the balance that must be struck between the free exchange of information and the protection of personal freedoms.

As any computer expert knows, one of the most complex technical problems is the place where the database will be stored. The United States, the Swedes say, have two-thirds of the bibliographical databases, particularly data of a scientific and technological nature. The French complaints are quite similar.

Most European countries look at it this way: from this concentration derives a great dependence, at least in certain sectors, such as in the medical profession and in chemical engineering. It is the semiconductor story all over again.

Supremacy in databases also plays a role in economic information. It is possible on the American networks to obtain very complete economic and financial data concerning the situation of most countries around the globe, using simulation models to help in comparison and evaluation.

Euronet has been chosen to fill this gap between American database technology and European availability. Originally projected for scientific and university data transfers, Euronet is now beefed up to reduce Europe's growing dependence on the established U.S. data bank services, whose only rival so far has been the European space agency's system in Frascati, Italy (and its service is limited and highly specialized).

If Euronet satisfies its sponsors' goals, Western Europe will eventually become a net exporter of scientific and technical information. The consequences would be a sharp reduction in the European revenue of Lockheed, SDC, and other U.S. data firms, whose combined European sales have been growing at the annual rate of 20%. (Lockheed Information Systems has the most at stake because it has gained 60 to 70% of the European market. Euronet starts operations after a common market outlay of $8.8 million. Spending of another $11.5 million is planned over the next three years to expand the system's capacity and promote its use. Euronet charges will be 20% below those of the U.S. concerns, partly because transmission costs will be lower—$2.70 an hour, compared with $12 an hour.)

Euronet will sell precisely the same kinds of information sold by U.S. data companies. These offerings include, for example, abstracts of papers on chemical and biological subjects, engineering data, patent information, and recent physics findings. U.S. concerns currently provide a far greater volume, about 80 separate databases compared with Euronet's initial 20 to 25. But Euronet's databases are expected to double within the first year.

Let's add that whatever is true between countries can also be true between poorer and richer regions within the same country, between cities, and even families within the same city—all the way to individuals. The Ten Commandments did not foresee this sort of challenge.

PERSONAL PRIVACY, PERSONAL FREEDOM

"Transborder" movement of data is a broad issue. But a much broader one stands at the personal, family, and state levels. Data involve at the same time:

- Personal privacy
- Personal freedom
- Wealth—and its understructure
- Know-how

The key issue is this: At this time, the transmission of data across a border—whichever border it may be—is characterized by two basic factors:

1. *Lack of government authority:* The police, the customs, and communications authorities are not endowed by law with the right to control the movement of data.

2. *Rapid expansion:* Although present transborder data transfer is a rather limited activity, it is also of great potential. This is the more true as data communications is due to play a significant role not only in international commerce and finance but also in the development of civilization down to the tribal level.

Data can be of a strategic nature and their dislocation can shake nations, organizations, and the family as well. Risks can be identified at more than one organizational level:

- Industrial espionage, and in general the transfer of R&D, banking, and business secrets

- Dependence—from the sciences to all aspects of the cultural life

- A possible tax drain of taxation by siphoning off information on benefits, salaries, and the like

- Vulnerability due to events foreign to this country but linked online, such as strikes, lockouts, and boycotts

- An amplification of destabilization movements in international commerce, particularly as the exchange of information brings into perspective the differences that exist in the international division of labor.

Technological progress both supports and is supported by the existence of media for transmitting information. Computer networks will allow increasingly rapid and massive data transfers at a highly reduced cost, at tariffs independent of distance and based on volume.

Videotex and electronic mail will surely accelerate data transmission and have a tremendous impact on both communications and sociology.

These are the background factors that make the need for harmony and coordination of legislatures deeply felt. In fact, discussions are in process within the OECD with the objective of bringing the American and European approaches closer and deciding on common rules.

These discussions have emphasized the great complexity involved in the elaboration of an international juridical instrument able to rule on the privacy of information. But what about the impact of information exchange and data banking on family life?

GLOSSARY

CCITT: International Telephone and Telegraph Consultative Committee. The members of this International Committee are the telephone company and telecommunication authorities of each country. The United States is represented through a special office of the State Department. Although CCITT has a consultative character, its "recommendations" in the field of telecommunications stand a good chance of becoming the international standard. A rather recent example is the X.25 recommendation for packet switching.

Circuit Switching: Circuit switching involves the establishment of a total path at all initiations; it is established by a special signaling which threads its way through different switching centers and is subject to the speed and code limitation of the weakest link.

Closed Information: Interactive videotext information which is intended for only a restricted user audience and calls for the use of a key (code) number for access: for example, confidential company information or text, data, and images of high current value sold to subscribers. This concept makes feasible closed user groups (CUGs).

Compunications: Compunications is the blending of voice and data communications with computers. It has been one of the vital links in telematics and can be put into three main types of use: replacing lengthy telephone calls, substituting for short memos, and introducing new communications ways that were not possible before.

Decision Support Systems: Computer-based algorithmic approaches, transparent to the user, which operate on text and databases to present to senior management (preferably in graphical form) the information needed for important decisions.

The trend in DSS will be to answer questions and work on problems as yet undefined, precisely within the unstructured information environment characterizing the executive level.

Editing Terminal: A terminal designed for use in projecting, preparing, or modifying viewdata pages. It is distinguished by its facilities for encoding alphanumeric, color, and graphics information. Online editing terminals operate directly to the viewdata-base computer. Offline (or local) editing terminals permit the local preparation, viewing, and modification of viewdata pages (both routing and information).

FDM: Frequency-division multiplexing, a transmission technology employing different frequency bands with a means at both ends of discriminating and recognizing the assigned frequencies.

Home Viewdata: The use of interactive videotex in the home environment. This can be assisted by home computers. Early examples include such applications as TV games and chess.

Information Provider: An organization that provides information for storage on a viewdata base. Typically, these are companies that rent space on the public utility system, which can be called up by viewdata subscribers, and will be charged by the organization running the interactive videotex service. After deduction of operation costs, the resulting revenue will be passed on to the information providers concerned.

In-House Systems: Viewdata systems operated by a single information provider on which to display his or her own text, data, and images (although nothing precludes that the information provider also rents space on his or her private system). In-house services can be supported in one location and in long-haul through leased or public telephone lines.

Message Switching: Because of store and forward capabilities, message switching requires no direct wire connection; it permits broadcast of messages; delayed delivery is feasible if the recipient is not available or if lines are busy; optimization is made in terms of throughput, but the network and its functions are applications-oriented.

Office Automation: Office automation is the process that integrates a wide variety of productivity-related products into computer-based systems for office work. Its perspectives range from the narrow view of bringing together copying, text, text handling, and distributing to the broad issue of providing the necessary understructure for data, text, image, and voice.

Open Information: Text, data, and images accessible to all viewdata subscribers. Hence public information is open information: for example, train and airline timetables, restaurant guides, general news, and so on.

Packet Switching: Like message switching, there is no direct wire connection, but the network and its facilities are applications independent. Packet switching (usually) employs minicomputers at the nodes, permits broadcast of messages, allows users to employ small or large bandwidth according to need; makes speed and code conversion feasible; dynamically establishes a route for each transmitted packet; and makes reliability reasonably high.

Prestel: A viewdata service operated by the British Post Office (BPO).

Public Utility Systems: Viewdata systems operated by organizations whose primary role is to provide a service capable of supporting a variety of information providers in displaying their information and a large number of users. Practically any telephone user can become a subscriber.

SDM: Space-division multiplexing, the eldest of the transmission technologies, associated with the physical metallic path which exists in a given switching connection.

TDM: Time-division multiplexing, a transmission technology employing time segments for channel separation, used primarily with wideband carriers and a large number of subscribers to be accommodated.

Telematics: Telematics is a synthesis of functions—the integration of voice communications with text, image, and data communications into a new system. In a broader sense, it also includes entertainment and the postal service.

Teletex: This system operates at normal voice-grade line speeds rather than the slower telex circuit line speeds, with a full alphanumeric character set. Although not yet directly related to viewdata and teletext, the name "teletex" can bring confusion if it is not clarified. This is the CCITT term for electronic mail.

Teletext: This is a *one-way* system. Basically, a broadcast message service using the TV signal as carrier and based on a standard character set which can be received and interpreted by a normal TV set adapted to that purpose. Text and graphics information is transmitted with the television signal. Approximately 250 pages can be transmitted in sequential and repetitive form at a rate of about four pages per second. At the receiving end, the domestic user can select the required page by means of a simple numeric keyboard, such as an adapted TV remote control unit. An index appears automatically on the user's TV screen when the teletext mode is selected. When the required page next occurs in the transmitted signal, it is retained in the receiver and displayed. The CCITT term is "broadcasting videotex."

Videotex: An interactive message service (two-way, send/receive) using a standard character set which can be controlled and received over voice-grade telephone lines and interpreted by a normal TV set, VDU, or other viewing device with adapted circuitry.

The adapted television set includes the line modem and an automatic dialer to establish the contact with the viewdata computer directly. Once contact has been made, an introductory page is sent back to the receiver.

The user then keys in his or her access number to the system and is accepted by viewdata. The central computer than sends a menu of alternative subjects for selection by the user.

A key feature of viewdata is this provision of text, data, and image in the form of multiple layers accessed via menus. It brings the user quickly to the required information, and it is user-friendly.

The CCITT term is "interactive videotext;" in England (where the service started) it is known as "Viewdata."

Videotex Charges: Charges for the system include the telephone line and a cost per page, based on the value of the information (higher charges for specialized data). There is no charge for menus, classified advertisements, and some of the infopages.

Videotex Features: These include the ability to restrict parts of the database to access via security keys and only from authorized terminals; automatic billing of the charges to the user; credits to the information providers; the possibility of making purchases via the viewdata system; and the facility to handle messages.

Videotex Page: A single frame (screenful) of information, containing 24 lines each of 40 characters of the standard viewdata character set. Normally only the middle 20 lines are free for information. The first and last lines are reserved for the viewdata system, and it is good to leave a buffer line on each side.

INDEX